THE GREAT LIVES SERIES

Great Lives biographies shed an exciting new light on the many dynamic men and women whose actions, visions, and dedication to an ideal have influenced the course of history. Their ambitions, dreams, successes and failures, the controversies they faced and the obstacles they overcame are the true stories behind these distinguished world leaders, explorers, and great Americans.

Other biographies in the Great Lives Series

CHRISTOPHER COLUMBUS: *The Intrepid Mariner*

THOMAS EDISON: *Inventing the Future*

JOHN GLENN: *Space Pioneer*

MIKHAIL GORBACHEV: *The Soviet Innovator*

JESSE JACKSON: *A Voice for Change*

JOHN F. KENNEDY: *Courage in Crisis*

MARTIN LUTHER KING: *Dreams for a Nation*

ABRAHAM LINCOLN: *The Freedom President*

SALLY RIDE: *Shooting for the Stars*

FRANKLIN D. ROOSEVELT: *The People's President*

HARRIET TUBMAN: *Call to Freedom*

ACKNOWLEDGMENT

A special thanks to educators Dr. Frank Moretti, Ph.D., Associate Headmaster of the Dalton School in New York City; Dr. Paul Mattingly, Ph.D., Professor of History at New York University; and Barbara Smith, M.S., Assistant Superintendent of the Los Angeles Unified School District, for their contributions to the Great Lives Series.

AMELIA EARHART

CHALLENGING THE SKIES

By Susan Sloate

FAWCETT COLUMBINE
NEW YORK

My deepest thanks and gratitude
to Mrs. Virginia Oualline and Mrs. Loretta Gragg
of The Ninety-Nines, Inc.,
for their gracious cooperation and assistance,
and to Mr. John Taylor
of the National Archives
for his extraordinary help
above and beyond the call of duty
in tracking down answers
to the Earhart mystery.

For middle-school readers

A Fawcett Columbine Book
Published by Ballantine Books

Produced by
The Jeffrey Weiss Group, Inc.
96 Morton Street
New York, New York 10014

Library of Congress Catalog Card Number: 89-90903

ISBN:0-449-90396-6

Cover design and illustration by Paul Davis

Manufactured in the United States of America

First Edition: February 1990

10 9 8 7 6 5 4 3 2

TABLE OF CONTENTS

The Ninety-Nines

This 1923 portrait of Amelia Earhart was used for her pilot's license.

1

Woman in the Clouds

THE SKY OVER Los Angeles, California, was blue and endless in October of 1922. One could see clearly for miles in any direction. It was a perfect day for an air show.

Rogers Air Field in Los Angeles was crowded with spectators. Stands had been set up on the field, and they were filled with people waiting to see daredevil pilots in small planes swoop around the sky. Airplanes were a new and dangerous mode of transportation, and air shows were a great way to interest the public in flying.

Among the crowd at the airfield were two young women and an older man. The man was Edwin Earhart, a Los Angeles attorney. The young women were his two daughters, Amelia and Muriel. Amelia, the older girl, had invited her father and sister to the air show.

As they reached the stands, Amelia produced a

pair of tickets and handed them to her father and sister. "These are for you," she told them. "I can't sit with you."

Edwin and Muriel were startled. What did Amelia mean? They bombarded her with questions as they settled into their seats, but Amelia answered none of them. She merely waved at them, smiling, and disappeared in the direction of the planes on the field.

Twenty-five-year-old Amelia was a tall, slender young woman with short, tousled blond hair, steady gray eyes, and a firm, straight nose. As usual, she was dressed in shades of her favorite color, yellow. When she smiled, a large space showed between her two front teeth.

Amelia had been interested in flying for a number of years. Just a year and a half before, early in 1921, she had actually learned how to fly. She was one of the first women in the United States to learn this dangerous and exciting skill.

Aviation — the production and flying of craft that are heavier than air — had begun in 1903, on a windy hill in Kitty Hawk, North Carolina. It was then that two brothers, Wilbur and Orville Wright, tested their homemade airplane. While Wilbur watched on the ground, Orville lay flat on the wing of the plane as it lifted into the air. The plane soared for just twelve seconds, and traveled a distance of only one hundred feet and twenty feet before landing in the grass.

But twelve seconds was all the world needed. Man could fly! Almost at once, others began to copy the Wright brothers' method of building planes. They, too, wanted to fly.

The first "flying machines" were a far cry from the sophisticated airplanes of today. They were small craft that ran on just one motor. Early planes had an open cockpit that could fit two people at most. They usually had two sets of wings, one above the other; these planes are known as biplanes. Planes with just one set of wings are called monoplanes.

For over ten years, aviation pushed forward in small, isolated steps. A builder would make a plane and fly it himself. Then someone else would build a plane, incorporating the first builder's ideas and adding his own. It was a long, slow process.

Aviation took a great leap forward when World War I broke out in Europe in 1914. For the first time, man fought not only on horseback, on foot, and on ships, but also in rolling tanks on the ground, and in the air. Lighter-than-air balloons with rigid frames called zeppelins dropped bombs over Europe. German designers engineered planes that could be used to take photographs from the sky. British airplane builder T.O.M. Sopwith designed the Sopwith Camel, a plane that could be used as a fighting craft.

Daring young pilots used the tiny planes to fight other daring pilots in the skies over Europe. At first the pilots used pistols in the air, but later engineers mounted machine guns in their planes. Combat between these fighter planes was called "dogfighting," and the pilots loved it. Each time they scored a "kill," or shot down a plane, they marked it on the side of their planes. When a pilot had scored five kills he was called an ace.

Aces were daring heroes, and they dressed the

part. They wore dark goggles, high black boots, worn leather jackets, and silk scarves tossed over their shoulders. They were romantic, and they knew it. To them, nothing made life more worth living than being able to jump into the cockpit of a little open plane and head for the skies!

Flying fighter planes was so thrilling that when the war ended in 1918 fliers did not want to return to the ground. Flying had changed their lives. They were reluctant to go back to the small towns of America they came from and resume work there. Instead, they decided to keep flying.

Many of these pilots became "barnstormers." These men owned small planes, which they flew into different American communities. In their silk scarves, leather jackets, and goggles, they were an impressive sight. Barnstormers would soar high in the air, showing their wide-eyed audience all the glory of flight. After a suitable demonstration they would invite spectators to fly with them — one at a time, and for fees of about five dollars for ten or fifteen minutes in the air. At night, barnstormers would tie down the planes near a barn in order to avoid damage to the craft in case of a storm. The next morning, they would be on their way to a new community full of new admirers.

Barnstormers also participated in air shows. They flew together, thrilling audiences with their dangerous stunts. "Wing walking," for instance, was a stunt in which one pilot flew the plane high above the ground while another walked on the wings. Or a barnstormer might attempt to move from one plane to another while the planes were both in the air, hovering almost wing to wing. This type of flying was exciting — and extremely dangerous. Not surprisingly, many barnstormers were killed in accidents in full view of their audiences.

Finally, in the 1920s, a board that would later become the Federal Aviation Administration (FAA) was created to regulate flying. This board sought to introduce flying to the public as an alternate means of transportation. Planes were faster than trains or automobiles, and the regulatory board wanted to assure the public that they were also very safe. Pilots now had to have licenses in order to fly. The board refused to permit barnstorming, which they considered too dangerous. They continued to allow air shows, however, in order to remind the public that flying was both fun and safe. They wanted very much to keep flying in the public eye.

At this time, flying was a rarity. There was no such thing as commercial aviation. No one flew as a passenger in order to get from one place to another. Flying was simply a thrilling indulgence, like driving a race car. No ordinary person would climb into a plane as a means of transportation.

That was how matters stood in October 1922. Few pilots held licenses, and even fewer were women. Amelia Earhart, who received her license in 1921, was one of the amazing women who took to the skies.

Amelia had become a familiar figure on the airfields around Los Angeles. Like many fliers, she wore the aviator's dramatic uniform: a brown leather coat that was scuffed and faded, tight-fitting breeches, a long white silk scarf, dark goggles, and knee-high black laced boots.

Amelia was at ease with the men at the airfields. She was not a society girl who was afraid to get her hands dirty. Amelia worked alongside the men, and she was interested in learning everything she could about airplanes. She was kind and friendly, interested in others, and possessed an amazing degree of courage.

Edwin Earhart and his daughter Muriel were surprised on that October day when Amelia left them in the stands at Rogers Air Field. Soon they heard the announcement that Miss Amelia Earhart would attempt to break the current women's altitude record. She would try to fly higher in the sky than any woman had ever flown before!

The Earharts strained their eyes as a small, single-engine airplane skimmed along the ground in front of them and lifted into the air. The plane climbed and climbed, until the Earharts and the other onlookers had to crane their necks and strain to see through the high clouds. Soon they could no longer see even the reflection of the sun on the plane's wings. All they could do was listen for the steady hum of the engine. As the hum grew fainter and fainter, Edwin looked apprehensively at his younger daughter. Was Amelia all right? Was the plane still flying safely?

Suddenly the plane reappeared out of the clouds. With the sun shining on its silver wings, it looped downward and landed neatly on the grass. Amelia, smiling and very composed, stepped out of the cockpit. The men on the field rushed over to the plane to read an altitude-measuring instrument that had been fastened to the cockpit. They gasped. Amelia had flown up to 14,000 feet to set a new altitude record for women!

This was the first of many records Amelia would set in her short but amazing career as a flier. As one of the most well-known pioneers of aviation, she would stand at the forefront of a group of courageous fliers who transformed public thinking about flying. She encouraged women to become a part of aviation by demonstrating again and again just how much she herself could achieve.

Amelia's desire to reach out to new boundaries was the motive behind everything she did. She refused to accept the limitations that restricted other people's lives. As a result, she pushed back boundaries for everyone. She knew that flying was unspeakably risky, but she chose to do it anyway. She did not let the fear of failure, or even death, deter her from exploring new possibilities. She knew that someone would be the first to move ahead in aviation, and she always wanted to be that someone.

2

Amelia's Childhood

THE GIRL WHO grew up to fly all over the world was born on July 24, 1897, in Atchison, Kansas, at the home of her maternal grandparents. She was named Amelia Mary Earhart, after her two grandmothers.

Amelia probably inherited a good deal of her curiosity and eagerness to explore from her mother, Amy Otis Earhart. Before her marriage, Amy Otis had ridden circus horses fearlessly, and she had been the first woman to climb to the top of Pike's Peak in Colorado. Yet Amy was also the daughter of a prominent judge, Alfred Otis, and she was accustomed to a very secure life-style.

Amelia's father, Edwin Stanton Earhart, was a young lawyer who had hoped to someday earn a seat on the United States Supreme Court. However, when he married Judge Otis's daughter in 1895, the judge let Edwin know that Amy was accustomed to the best of everything. Judge Otis felt that the challenges his son-in-law would encounter as an ambitious judge would make it difficult for him to care for his family. Judge Otis told Edwin that, above all else, he expected him to provide a comfortable home for Amy.

So Edwin swallowed his dreams of a seat on the Supreme Court and instead went to work as an attorney for the railroads in Kansas City, Kansas. At that time, this was the best-paying legal work available. In his spare time, Edwin fiddled with various kinds of inventions, trying to create something that would bring even greater financial security to his family.

In 1897, two years after Edwin and Amy were married, Amelia Mary was born. A second daughter, Grace Muriel, was born on December 29, 1899. From the beginning, the two little girls were best friends and playmates. Muriel called her big sister "Meelie," and Amelia called Muriel "Pidge."

The girls had a progressive childhood, by the standards of the day. Neither of their parents believed in the restrictions on dress and behavior that many girls labored under all of their lives. They also raised Amelia and Muriel to believe that they were the equals of boys. Amy permitted her daughters to dress in bloomers rather than dresses, so they could feel comfortable while fishing, playing baseball, and exploring by the river.

Neither Amelia nor Muriel were interested in the games other little girls of the early 1900s were encouraged to play, such as tea parties on the lawn and playing house. They much preferred active outdoor games. The neighbors objected to their "boylike" behavior, but the girls' parents allowed them to spend their time as they chose.

So Amelia and Muriel romped and played, sometimes with neighbors' children, just as often by themselves. Amelia particularly liked to pretend that she owned a beautiful Arabian stallion, which she liked to "gallop" all over the backyard. She and Muriel collected specimens of bugs and lizards to study. And like children everywhere, they gloried in family picnics that featured homemade ice cream.

There was something that the growing Amelia found even more fascinating than these pastimes, however: machines that generated speed. In 1904, the entire Earhart family spent a week in St. Louis, Missouri, at the World's Fair. Amelia, who was then not quite seven years old, was captivated by the sights, sounds, and smells of the fair. She was especially thrilled about one ride, which featured small cars speeding around an aerial track.

However, Amelia's mother thought the small cars were too dangerous for her young daughters. Amy refused to allow Amelia or Muriel to ride in them, though the girls were permitted to ride an elephant. So Amelia swallowed her disappointment and contented herself with other rides, such as the Ferris wheel, which was large and impressive but did not move very fast.

The sight of those speeding cars stayed in Amelia's mind. When she and the family returned to Kansas City, Amelia decided to construct her own roller coaster in the backyard of her home. She chose the toolshed as the starting point of the coaster chute, and she persuaded Ralphie Martin, a neighboring boy who owned a tool set, to come over and help.

But Amelia and Ralphie ran into technical difficulties. Amelia asked her uncle, Carl Otis, how to manage the construction, and he explained the whole process to her. She labored all day with Muriel and Ralphie to set up the roller coaster. They built a long chute of wood, and constructed a wooden car with wheels that could roll down the chute.

Amelia, as the inventor and master builder of the coaster, decided to take the first ride when the project was completed. Muriel held the little homemade car while her sister climbed in. Then Amelia shrieked, "Let me go!"

The car rolled down the track at a terrific speed. As it reached the bottom, it flew off the chute altogether, and Amelia landed on the grass. The accident resulted in a bruised lip and a torn dress, but the young inventor barely noticed these injuries. She jumped up, eyes shining, and exclaimed to her sister, "Oh, Pidge, it's just like flying!"

Then Amelia started to figure out how to improve the ride. Unfortunately, their mother was horrified when she found out what the girls were doing. Amy believed that the roller coaster was extremely dangerous, and she insisted that the girls dismantle it immediately. She tried to make it up to the disappointed children by giving them a lawn swing, but that wasn't nearly as much fun.

At the same time that Amelia's love for speed and motion was growing, she was also learning important lessons in character. She developed great compassion and the conviction to stand up for her beliefs. Amelia may not have known it then, but these qualities would remain very important to her for the rest of her life.

When Amelia was just eight years old, she took on her first cause. Her next-door neighbor refused to build a stable for his mare, Nellie, who suffered in the summer heat and was bothered by the buzzing flies. Amelia's heart went out to the horse. She persuaded her sister to help her aid Nellie. They could not move the unhappy mare, since she did not belong to them. But when they heard Nellie starting to kick and neigh, they ran over to comfort her with warm pats and handfuls of clover. Nellie always calmed down at once when the girls did this.

When Nellie was killed in an accident caused by her owner, an indignant Amelia refused her mother's instructions to take the bedridden owner a piece of cake. She told her mother that he was so horrid she would not take him anything! In the early 1900s, a child who disobeyed her mother's orders could expect a spanking, if not worse. But young Amelia stood her ground. She would not give in when she felt she was doing the right thing.

This refusal to back down from a just cause continued to flare up as Amelia grew older, particularly after she began to read. She and Muriel both could spell out their favorite books and read them aloud before they were five years old. Together they began to read the Oliver Optic success stories for boys. As she got older, Amelia especially enjoyed novels by

Charles Dickens and adventure books such as *The Count of Monte Cristo* and *The Queen's Necklace*. Tales of brave men who fought for causes they believed in fanned the flames of her child's imagination. She imitated their admirable dedication in her own life by standing up for what she felt was right. No issue was too small or unimportant for her. Throughout her life, reading would be a source of both pleasure and inspiration to the idealistic Amelia.

Amelia's personality was also shaped by another factor: her family situation. Edwin Earhart began to realize that his personal dreams of glory would not be fulfilled because he had a young family to support. This realization hurt him deeply, and in his disappointment he turned to alcohol.

Edwin's work at the railroad office soon began to suffer. As Amelia and Muriel grew up, Edwin drank more and more often. The girls dreaded seeing him this way. Sober, their father was their favorite playmate. But when he was drunk, he turned into a man they didn't recognize. Tension mounted in the house as the years passed and Edwin's drinking disrupted family activities. Finally, he was fired from his job.

Edwin tried to pull himself together and find new employment. But his reputation as an alcoholic was too well known in Kansas City. No one wanted a heavy drinker working on important legal documents. He was forced to seek work in another city.

Finally, in early 1907, when Amelia was nine, Edwin was offered a job in the Rock Island Railroad's claims department in Des Moines, Iowa. He decided that although the job would mean a move for the entire family, it would be worthwhile to accept it. He and Amy left Kansas City for Des Moines to look for a house, while Amelia and Muriel stayed with their grandparents in Atchison, Kansas, in the house where Amelia was born.

In the summer of 1908, the girls joined their parents in Iowa. Des Moines was vastly different from Kansas City. Here they were no longer "the Judge's granddaughters," whom everyone knew. There was no more private school, no more private park to play in. Now, instead of a servant, it was Amy who fixed the girls' breakfast.

Iowa was an eye-opening world to Amelia. Amelia and Muriel adapted themselves to new schools and new neighbors. They made new friends. They were introduced to the magazine club that existed on their block, and they loved it. The children bought one copy of all the magazines that came out every month. Each child read the magazines, then passed them to his neighbor, until everyone had read all of them.

The Earharts encouraged the girls to enjoy music in their new town. Concerts were held at Drake University near their home. Amelia, like her father, had a natural ear for music. She easily picked up the fundamentals of playing certain instruments, such as the piano and banjo. She grew to love classical music, and listened to it whenever possible.

It was in 1908 that eleven-year-old Amelia first saw the flying machine known as the airplane. Her father took her and Muriel to the Iowa State Fair, where they enjoyed the many exciting booths and rides. In a large open area to the side of the fairgrounds, Edwin Earhart spotted an airplane sitting on the ground. The man standing next to it was the pilot, who took up passengers for a ride in the air. At that time, airplanes were rare, exotic things. For most of the people who rode in the plane, it was their first flight.

Edwin pointed out the large silver-colored contraption, which Amelia and Muriel had only seen before in magazine pictures. He told them a little about the Wright brothers and the first flying machine, which had been invented just five years earlier. Surprisingly enough, Amelia was disappointed at what she saw. She thought it looked like an orange crate. In fact, she was so unimpressed that she did not even ask her father for permission to go up with the pilot as a passenger. Little did she know that those flying contraptions would one day become the greatest passion of her life!

Meanwhile, her father's drinking continued to affect Amelia deeply. Edwin was eventually fired from his job at the Rock Island Railroad. Naturally, he sought a new job in legal claims, but although he wrote to railroads all over the country, it was difficult to find a new position. Finally the Great Northern Railway in St. Paul, Minnesota, offered him a job as a minor clerk. It was the best Edwin could do. Again, the family prepared to move.

The process of breaking ties and moving on became very familiar to Amelia and Muriel as the years went on. Amelia learned to adapt quickly to new situations. She became increasingly more independent, and helped her mother to run the household.

She managed to maintain excellent grades in school, no matter what school she was in. But Amelia also maintained a certain distance from her classmates. She had learned from experience that it hurt too much to develop close friends whom she would only have to leave when the Earharts moved again, as they always did.

Amelia's experience in dealing with her father gave her both great personal strength and compassion for other alcoholics. She would encounter many people with drinking problems during her life, and she was always especially gentle with them.

From St. Paul, the Earharts moved to Springfield, Missouri. Edwin had been promised a position there with the Chicago, Burlington, and Quincy Railroad. When the family arrived, however, Edwin found that a mistake had been made. There was no vacancy in the office. The Earharts had moved all the way to Missouri and there wasn't even a job!

Edwin's wife, Amy, had become frustrated with their life-style. This time she spoke up. Moving around like gypsies was bad for the girls, Amy said. She wanted them to live in a more stable environment. Furthermore, she had taken the situation into her own hands, and had found a place for them to settle down. The Earharts had close friends in Chicago, the Shedd family. The Shedds had offered to open their home to Amy and the girls. Amy told Edwin that she was going to take Amelia and Muriel to live there. They agreed that Edwin would remain in Springfield for a while, then try to open a law office in Kansas City. Once he was established in Kansas City, Amy and the girls would return. The separation of husband and wife would only be temporary.

Once again Amelia packed her things and moved. For the next year, she and Muriel and their mother lived first with the Shedds, and then on their own in a small apartment near Chicago, in Hyde Park, Illinois. By this time Amelia was about fifteen years old, and Muriel about thirteen. The girls entered high school in Hyde Park. Amelia's grades were excellent, but an issue arose that made her very unpopular with her classmates.

There was an English teacher at the high school whom Amelia felt was incompetent. Amelia led a crusade against her, requesting that the teacher be replaced by one who could instruct the students properly. But Amelia's classmates jeered at her. They didn't want a competent teacher; they wanted a class period in which they did nothing but have a good time. When Amelia asked for support from them, they just laughed. It is not known whether or not the teacher was ever fired, but one thing is clear: Despite the other students' criticisms, Amelia didn't back down from her cause.

Lack of popularity was the price Amelia paid for her independence and sense of justice. Her main concern was always to do what she felt was right, even if the other students didn't agree with her. As a result, she became something of an outcast at school. In June 1916, when Amelia was eighteen, she graduated from Hyde Park High School but did not attend the commencement ceremonies. Her picture in the school yearbook was captioned, "The girl in brown who walks alone."

Amelia, Muriel, and their mother united once more with Edwin in Kansas City that summer. Unfortunately, it was a brief reunion. Edwin was still drinking, and Amy and the girls did not feel comfortable around him. Finally, Edwin set off for Los Angeles to try to end his drinking for good.

Amy and Muriel returned to Chicago. Amelia, who was now nineteen years old, did not go with them. Although many young women did not pursue any education after high school, Amelia wanted to do so. In fact, it was her mother who encouraged her to continue studying, and allowed her to choose her own school as well.

Amelia decided to attend school in a city away from her mother and sister. She asked to be sent to the Ogontz School, a college preparatory institution in Philadelphia, Pennsylvania. She began studies there in the fall of 1916. Amelia quickly became part of the school's mandolin society, as well as a member of one of the three secret societies, or social clubs, on campus.

It was typical of Amelia to protest when she found out that there were girls who did not belong to any of the societies. She knew how it felt to be an outsider, how painful rejection could be. Amelia set out to correct this situation for the other girls. She asked the school's headmistress, Abby A. Sutherland, if the Ogontz School would permit the creation of a fourth secret society, to which these girls could belong.

There is no record of whether or not Amelia succeeded, but the enterprise obviously stemmed from her desire to help others, and displayed her courage in exploring new areas. It is unlikely that any of the other students would have thought of speaking up on behalf of others who were less popular. Amelia did not allow that to influence her actions, however. This is just one of many examples of how Amelia was always at the forefront of change and progress, even before she began flying.

During her years at the Ogontz School, Amelia also objected to the limited range of subjects the girls were allowed to study. Certain topics, such as women's rights, were considered "unsuitable" for young women, and were not included in the Ogontz curriculum. Amelia believed in free discussion and exploration of any topic in which a student was interested. In her mind, no topic should be banned. No book should be off-limits.

She herself read about as many different subjects as she could. In later years, Amelia's first flying instructor and close friend, Neta Snook, always saw Amelia with a book in hand. She read poetry, history, and even a large volume on Islamic religion — just to find out what it was all about.

These were times in which women were not expected to play an active role in national or world affairs. It was rare to find a woman who would dare to reach beyond the boundaries of "ladylike" behavior. Ladies were expected to stay home, tend to their husbands, their children, and their sewing, and remain ignorant of world affairs. Ladies did not work outside their homes. They did not vote, and they dressed in only the most modest clothing.

Before long, world events changed all that.

By 1917, World War I was already raging in Europe. In the spring of that year the United States finally entered the war, and most men left to fight in Europe. Suddenly American women were needed in all kinds of new capacities. They worked as nurses, mechanics, and ambulance drivers, among other things.

Many women responded enthusiastically to the call for help by plunging wholeheartedly into the business of war. Amelia admired these women who gladly shouldered the burdens men had carried for centuries. Something in her adventuresome and independent nature drove her to be one of them. She kept a scrapbook on these amazing women. She cut out articles and pictures wherever she could find them to paste into her book.

Amelia would become one of those amazing women sooner than she thought. All it took was one visit to her sister.

3

Lessons in War

LIKE MANY AMERICANS, Amelia was unprepared for the scope and horror of the first World War. World War I was the first modern war to engage all the major military powers of the world. The Allied Powers — the United States, England, France, Russia, and Italy — fought against the Central Powers — Germany, Austria-Hungary, Bulgaria, and Turkey. It was a war fought largely over the acquisition of land.

There had been no full-scale war in Europe for about one hundred years before World War I broke out. But for decades political, economic, and military changes in European countries had caused many tensions. Unrest seethed beneath the surface, waiting for an excuse to erupt. The assassination of Archduke Franz Ferdinand of Austria in June 1914 provided that excuse. War soon followed.

The United States did not declare war right away. It

preferred to remain as far removed from the conflict as possible. However, when a British civilian ship, the *Lusitania*, carrying U.S. citizens on board, was sunk at sea by German submarines in 1917, America could no longer remain neutral. The United States declared war on Germany and entered the conflict on the side of the Allies.

For the Christmas holidays of 1917, Amelia visited her mother and Muriel at St. Margaret's College in the city of Toronto, where Muriel was attending school. Canada had entered the war several years ahead of America. Some Canadian veterans were beginning to return home at the same time as the fresh-faced American soldiers, the "doughboys," as they were referred to, were setting out for Europe.

During Amelia's visit to Ontario, the Earhart sisters often went strolling in the streets. There, Amelia saw firsthand the disastrous effects the war had had on those who fought it. All around them the sisters saw soldiers on crutches, blinded, or with missing limbs. Since the United States had just entered the war, no such sights yet existed in the United States.

Amelia's compassionate heart went out to these men. She became determined to help them in any way she could. After a brief investigation of hospitals and nursing facilities in Canada, Amelia wrote to her mother and to the headmistress of the Ogontz School. She requested that she be permitted to leave the school in order to become a volunteer nurse's aide in Canada. With her mother's very reluctant permission, Amelia took a concentrated

Red Cross course in nursing and joined the staff of Spadina Military Hospital, in Ontario, as a nurse's aide.

Here, Amelia blossomed. She loved being able to work in a place where she could truly be of service. She became popular with the wounded soldiers, and worked hard to provide good care. Amelia even convinced the hospital's dietitian to change the menus in order to give the men tastier meals. She managed to get their turnips and parsnips changed to tomatoes, and their rice pudding eliminated in favor of blancmange, a tasty dessert that was popular then. The men cheered her for it.

In her spare time away from the hospital, Amelia and Muriel rode horses at a nearby stable. They were often joined by officers of the Royal Flying Corps who were no longer hospitalized. These officers had flown in "dogfights" over Europe, engaging in fierce battles with enemy fliers. They had many tales to tell of courage and fearlessness. Amelia listened, fascinated.

Courage was by no means limited to the Canadian fliers, however. Amelia herself soon proved once again to be a brave woman unafraid of challenges.

There was an especially wild horse at the stables, aptly nicknamed Dynamite, who refused to allow a rider on his back. Amelia tamed the horse single-handedly. She visited Dynamite every day with treats such as apple bits and sugar. She talked soothingly to him. In only a month, the horse permitted Amelia and others to ride on his back. The stable grooms were very impressed, and refused to charge Amelia for her riding time because they said she had saved Dynamite for them.

The officers, too, took notice of Amelia's courage and camaraderie with Dynamite. One officer, Captain Spaulding, told her she rode her horse the way he flew his airplane. He invited her to see him fly in an air show at nearby Armor Heights the following week. Amelia agreed to attend.

Amelia brought another young nurse's aide with her to the airfield. The two women watched as daredevil pilots dove and looped in the air. One particularly daring pilot swooped down on them from the clouds, heading straight toward the crowd. The spectators promptly scattered, screaming in fear.

Dressed in their bulky nursing uniforms, Amelia and her companion were not able to move as fast as the other spectators. Instead, Amelia stood her ground, watching as the little red plane dove out of the sky toward her. She knew that if something went wrong with the controls of the plane, she and the pilot would both be killed. Still, she remained calm, looking up at the amazing plane. It was a momentous experience for her.

Years later Amelia said that as the plane went past her she thought it said something to her. It was the beginning of her romance with flying.

4

Taking to the Skies

WHEN THE WAR in Europe ended, in November of 1918, Amelia was recovering from a bout of pneumonia. At that time she was still not quite ready to pursue flying, though her fascination had been sparked at the air show in Armor Heights. Since she was no longer needed as a nurse's aide, she was faced with an important decision: What should she do with herself? Amelia wasn't sure of the answer to that question.

Twenty-two-year-old Amelia decided not to resume her studies at the Ogontz School after the war. Having been deeply affected by her nursing experience in Toronto, she decided instead that she wanted to become a doctor. In the fall of 1919, she enrolled in premedical courses at Columbia University in New York City.

But Amelia failed to find the fulfillment she sought in her medical studies. She had an aptitude for science, as well as an intuitive understanding of her subjects, but she was not as engrossed in her work as she had been while nursing in Canada. She began to believe she had no real affinity for medicine.

By the time Amelia completed her first year at Columbia, in the spring of 1920, her mother had rejoined her father in Los Angeles. Edwin had become a member of the Christian Science Church, an organization founded by Mary Baker Eddy. Christian Scientists believed in maintaining good health through prayer. Members of the Church refused to take any kind of medicine or see doctors. Instead, they preferred to cure themselves through contact with God. As a result of his joining the Church, Edwin had stopped drinking completely. He now had a successful law practice in Los Angeles.

In the summer of 1920, a disillusioned Amelia went home to her parents in Los Angeles. She did not plan to return to Columbia in the fall. At the age of twenty-three, she had done many different things, but nothing felt right to her as her life's work.

The Earharts lived in a spacious house on Fourth Street. They rented out some rooms in the house, and one of their boarders was a young engineer named Sam Chapman. Soon after arriving in Los Angeles, Amelia met Sam. The two young people liked each other immediately. They both enjoyed music and art, and Sam began to escort Amelia to various lectures and functions around the city.

Gradually, Sam became Amelia's first boyfriend. In romance, as in every other area of her life, Amelia was sensible and patient. She was content to see Sam as a good friend and spend time with him on a friendly basis. She was not looking for too much romance. She wasn't ready to get married. But she *was* ready for a complete change in her life.

That change began when Edwin took her to an air show in Los Angeles. Amelia had never forgotten the air show she attended in Armor Heights, and she was delighted to accompany her father on the outing. There her deeper feelings about aviation finally surfaced. Seeing the planes high in the sky, Amelia knew beyond a shadow of a doubt that she wanted to learn how to fly. Finally, she had found what she had been looking for all her life.

After the air show, Amelia and her father approached a young woman with flaming red hair who was dressed in a flier's leather coat and boots. The woman was Neta Snook, one of the first female fliers in America. In 1920, Neta had her own air school in Los Angeles where she trained student fliers and gave rides in her Curtiss Canuck airplane to the curious people who wanted to experience the glories of flight. Neta was only a year older than Amelia, but she was an experienced pilot and flight instructor. She knew immediately that Amelia would be an attentive student.

Amelia began lessons with Neta, but they were very expensive. Neta charged the same rate for all her students: one dollar per minute for every minute her plane was in the air during a lesson. Amelia did not have the necessary funds to cover the fee. Her father could not afford to help her out, either. So Amelia went to work to earn the money. She took two jobs, one at the telephone company, and another in the darkroom at a photographer's studio. Her job in the darkroom was to print photographs from negatives, and she worked at it whenever she was away from the telephone company or the airfield.

Neta taught Amelia in her own Curtiss Canuck biplane. Like all planes of that time, the Canuck had an open cockpit. While flying, the pilot and passengers were exposed to all the elements. It was impossible to fly without wearing goggles to protect one's eyes from the wind and the weather. A passenger reaching a hand outside the plane might feel as though his arm were being blown off. To Amelia, the experience was electrifying!

Some of the young women who flew with Neta thought it would be glamorous. They arrived at the airfield in silky dresses and high-heeled pumps. After the first few tries they discovered that flying an airplane was serious work, and that sometimes it was very dirty. Once they realized this, most of the young ladies left immediately. But Amelia was different. She learned from the beginning to dress sensibly, and she always remembered to account for the wind and the weather. Most early women fliers wore riding breeches and high boots. Amelia gladly adopted the uniform.

Amelia was also very interested in engines and how they functioned, and worked along with the men. Often, when it was time to go home, she would be streaked with oil and grime, but it didn't bother her. Amelia realized that the most important thing was not keeping spotlessly clean, but learning to fly.

Amelia encountered one particular problem that

The Ninety-Nines/Albert Bresnik

Amelia first saw an airplane at age eleven while attending an air show. Ironically, she was not very impressed with the new "flying machines," likening them to orange crates. Years later, however, she became entranced with flight and sought to make it her life's pursuit.

most pilots do not experience. Whenever she flew, her sinuses, or nasal passages, became inflamed. The thinner air at higher altitudes caused her blinding pain. She could not take medication for the pain because she feared it might impair her senses while she flew. The remedies she tried on the ground were useless. Amelia chose to brave the pain. Learning to fly was too crucial to her entire existence to let something like this interfere.

Slowly, Amelia began to build up her hours of flying experience, a process known as "logging flight time." As a beginner, she could not fly in bad weather, or in winds that blew in the wrong direction. She refused to fly alone, or "solo," until she felt completely qualified to handle the plane in any kind of emergency.

She was also very passionate about the idea of owning her own plane. Her father could not afford such a purchase, but would not have helped her to buy a plane even if he had enough money to do so. Edwin was afraid for Amelia's safety. Even though he was very proud of her independent thinking and her accomplishments, he knew aviation was still a very dangerous pursuit and worried about his daughter's enthusiasm for flying.

It was Amelia's mother and sister who helped make her dream a reality. Amy had come into a small inheritance from her parents. She used some of the money, and Muriel added her savings, to help Amelia buy a Kinner Airster. The Airster was a much smaller plane than Neta's Curtiss Canuck, but it was, nonetheless, Amelia's very own.

On July 24, 1922, the day of her twenty-fifth birthday, Amelia took possession of the sporty biplane, which was yellow, her favorite color. From then on, Neta taught her to fly in the Canary. The wingspan of Amelia's plane was much smaller than that of the Canuck's, and it was also less stable. But Amelia was very determined to master flying her plane, and she continued her training doggedly.

Now that she owned her own plane, Amelia insisted on looking even more like a pilot. She bought a leather coat like the one Neta wore. But the coat displeased her because it looked so new and shiny, not at all the way she imagined a pilot's coat should look. For several days she wore the coat around her home and in the garden hoping to wear off the shine of the leather. She was so desperate to give it an "aged" look that one night she even slept in it!

Not all of Amelia's training was pleasant. Only a few days after she first went up in the Airster, Amelia experienced the very dangerous side of flying. As she was returning to Kinner Field from another airfield, Amelia could not coax her plane to rise above a circle of eucalyptus trees at the end of the runway. When she pulled the nose of the Airster up to try to get above the trees, the plane stalled. The engines had stopped, which meant that the airplane had lost power completely and could no longer stay in the air.

The plane smashed into the ground, breaking its propeller and damaging its landing gear. Luckily, neither Neta nor Amelia were injured. When Neta turned to check that Amelia was all right, she found her student calmly powdering her nose. Neta was astonished, but Amelia said simply, "We have to look nice when the reporters come."

While Amelia was still taking lessons, Neta became engaged to be married. Once she married, Neta would no longer fly *or* teach. Amelia had not yet flown solo, and she wanted desperately to continue flying lessons. She decided to go on learning with an instructor who had taught flying in the army. His name was John "Monty" Montijo.

Over the course of the following year, Amelia became an extremely competent flyer. Still, she insisted on knowing acrobatic maneuvers before going up on her own. She spent ten hours on acrobatic flight instruction with Monty before she finally felt ready for the big step. On the day when Monty finally sent Amelia up alone, her sister, Muriel, was standing on the field watching proudly. Amelia had finally soloed! She was a pilot! A few months later, she achieved her first record by setting the new women's altitude mark.

Amelia's friend Sam Chapman was not very happy about her pursuit of flying. He felt it was dangerous, especially for a woman. But he was easygoing and quiet, and he tried not to interfere. Even if he had, Amelia would probably have paid little attention to his protests. She was doing what she wanted to do, and nothing was going to interfere with that.

Unfortunately, Amelia's triumph in the air was not mirrored at home. The Earharts' marriage was not as happy as their daughters had believed it to be. By the summer of 1924, when Amelia was twenty-seven, her parents could no longer live together. When Edwin Earhart filed for divorce, Amy did not object. Edwin decided to remain in Los Angeles, but Amy and Amelia drove east. Muriel was teaching junior high school in Medford, Massachusetts, a suburb of Boston. Amy and Amelia decided to join her there. With no money to spare for a car, Amelia decided, regretfully, to sell her beloved Kinner Airster and use the money to buy a bright yellow Kissel touring car. She and Amy set off for Massachusetts in early 1926.

The family settled in Medford. Since Amelia no longer had a plane, she gave up flying temporarily. It was a great disappointment to her, but her spirits soon lifted when she underwent successful nasal surgery to cure her sinus trouble. If she could ever afford to buy another plane, she would be able to fly without pain.

But flying would have to wait. Amelia, Amy, and Muriel had little money, and the most important thing was to find a job. For a while, Amelia taught English to foreign students in the University of Massachusetts Extension program. It was exhausting work, however. Frequently she had to drive thirty miles to reach a student, a journey that took two hours or more on the rough roads which were common at that time. Amelia often arrived at her destination exhausted. To make matters worse, the pay was very low. She was given a transportation allowance, but it hardly covered her traveling expenses.

Amelia needed a better situation, and she found it at Denison House — a neighborhood settlement in Boston inhabited mainly by Syrian, Irish, and Chinese immigrants. Amelia started out teaching English there, but her duties quickly expanded. Soon she was in charge of keeping the records of Denison's household expenses, and was also doing outside social work as a visiting nurse.

The children of the Denison House immigrants adored Amelia's yellow car, and she often took them for rides around Medford. But soon it became obvious to Amelia that she would be able to accomplish more work if she moved into Denison House. She did so, and thus became a resident worker there.

This disappointed Sam Chapman. He had followed Amelia to Massachusetts from Los Angeles. He was ready to marry, and he wanted to marry Amelia. But Amelia was not moved by Sam's plea of love. She did not want to tie herself down in a marriage that would require her to give up the activities she cherished. She would always need the freedom to do as she pleased, and Sam could not accept that in a wife. As gently as she could, Amelia turned down his marriage proposal. They could always remain friends, she told him, but she could never contemplate becoming the traditional homebound wife Sam envisioned. She was too engrossed in her day-to-day activities at Denison House. She also had other plans.

In the back of her mind, Amelia deeply missed flying. She loved working at Denison House, but the urge to fly proved too strong to resist. Amelia began to spend her days off at the Dennison Airport, outside Boston.

Bert Kinner, the man who had designed and built Amelia's first airplane, the Kinner Airster, had also come east from California. He was hoping to establish a sales agency for his airplane company. Kinner respected Amelia's abilities as a pilot. After all, she had set the women's altitude record of 14,000 feet in one of his planes some five years earlier, in 1922. Kinner asked Amelia to demonstrate his planes for prospective customers. In exchange, she could use the plane for her own flying. Amelia agreed enthusiastically. She was back in the skies!

It was at Dennison Airport that people began to notice an odd circumstance: Although men spoke roughly among themselves at the airfield, no one used rough language around Amelia. No one treated her with anything less than dignity and respect. If a man was telling an off-color story and Amelia walked into the hangar, he stopped speaking at once. If he didn't, someone kicked him in the shin or elbowed him in the ribs until he did. Amelia commanded respect — though she never visibly said or did anything to encourage it. It was simply part of her being.

Amelia felt very much at home at Dennison as she began to fly again. She enjoyed talking about engines and wings and asked questions of anyone who would answer her. She was always down-to-earth, straightforward, eager to learn whatever she could, and the first to admit that she didn't know everything.

By this time, too, Amelia was thirty years old. She had fashioned a life for herself that was about as perfect as she could wish it to be. Then, suddenly, world events intervened. Overnight, the life of the young social worker from Denison House took on the fairy-tale flavor of Cinderella.

5

Across the Atlantic

AMELIA LOVED EVERY moment of her own flying, and she could not help but be caught up in the excitement of other fliers' accomplishments as well. In 1927, there was one other flier in particular who moved aviation a giant step forward as the world watched and waited.

In May 1927, a shy, gangling young man named Charles Lindbergh climbed into the cockpit of his plane, *The Spirit of St. Louis*. Other men had crossed the Atlantic Ocean in airplanes before, but none of them had been alone. Lindbergh intended to fly solo across the Atlantic Ocean, from New York to Paris.

All alone, battling high winds and exhaustion, feeling disoriented in the darkness, Lindbergh fought his way across the ocean. After many hours, his weary eyes caught sight of the lights of a city beckoning in the distance. He checked his instruments and his map. It had to be Paris, it just had to be.

As Lindbergh set the little plane down on the airfield, his eyes gritty with fatigue, thousands of people swarmed around him, shouting and cheering. The police tried to hold them back, but the adoring crowd broke through barricades to reach the tired hero. They carried him on their shoulders, singing in jubilation. Lindbergh was startled. All the attention made him realize that he was no longer just another pilot. The "Lone Eagle," as he was nicknamed, had stepped from obscurity to fame overnight. He was suddenly the world's most beloved celebrity.

At Denison House, Amelia and the other workers were dazzled by the courage and perseverance that this quiet, stubborn pilot had shown in accomplishing what no one before him had dared. Amelia read everything the newspapers wrote about Lindbergh, which was a considerable amount. When "Lindy" returned to New York, he was met by cheering crowds and was given the traditional greeting for heroes — a ticker-tape parade down Broadway.

Immediately after his return, Lindbergh signed a contract with a promoter named George Palmer Putnam. Putnam was the grandson of his namesake, the publisher G.P. Putnam, and he arranged for "Lindy" to write a book about his life. George Palmer Putnam would soon play a pivotal role in the life of another aviator — a girl living in Denison House and dreaming about flights of her own.

Because of the enormous publicity generated by Lindbergh's successful crossing, a society woman, the Honorable Mrs. Frederick Guest, decided to purchase a plane. Mrs. Guest, who had been Amy Phipps of Pittsburgh, Pennsylvania, before her marriage, now lived in England. She had been a flier herself for a number of years. Now Mrs. Guest declared she would become the first *woman* to cross the Atlantic Ocean.

Other women had tried. At least four separate attempts had been made. In each attempt, the would-be pioneer had perished, along with her male flight crew. After Lindbergh's flight, however, Mrs. Guest was determined that a woman would successfully cross the Atlantic from the United States to England. She believed it would be an excellent gesture of goodwill between the two countries and that it would help to cement relations between them.

Mrs. Guest's family was opposed to her plan. Reluctantly, she gave up her plans to take the pioneer spot for herself, and decided to allow another woman to fly in her place. George Palmer Putnam and his associate, Captain Hilton H. Railey, were entrusted with the job of finding a suitable girl for the flight.

The girl they were looking for had to be very special. She had to be a licensed pilot. She had to be as appealing to the public as Lindbergh was. She had to agree to certain stipulations and to all publicity concerning the flight. She had to be as polished, educated, charming, and feminine as possible. If she was pretty, that wouldn't hurt either.

Putnam and Railey asked for suggestions among their contacts in aviation. One of the people they asked was Commander Richard E. Byrd, the author of *Skyward*, a book about his own flying experiences. The Commander remembered hearing of a social worker who flew at Dennison Airport on the outskirts of Boston. Her name was Amelia Earhart. Would Putnam and Railey be interested in interviewing her?

They were. Amelia picked up the telephone one day at Denison House, and the voice of Captain Railey, acting on Mrs. Guest's behalf, asked if she would be interested in a hazardous expedition across the Atlantic Ocean. Did she want to consider discussing such a flight? Amelia's answer was simple and direct. "Yes. How could I refuse such a shining adventure?"

Captain Railey made Amelia promise to keep the entire project secret. She took a two-week leave of absence from Denison House to travel to New York. There, she met with three men who were acting on Mrs. Guest's behalf: David T. Layman, Mrs. Guest's attorney; John S. Phipps, Mrs. Guest's brother; and George Putnam.

The interview focused not only on Amelia's experience as a pilot, but on many other things as well. The three men were interested in her personality. Was she interesting? Was she well educated? Did she speak well? Would she look good in the public eye? The answer to each of those questions must have been yes. Two days after Amelia returned to Boston, she received a note from Mrs. Guest informing her that she had been chosen to captain the Atlantic flight in an airplane called the *Friendship*.

Along with the note was a contract from Mrs. Guest's lawyer which specified Amelia's responsibilities as captain of the Atlantic flight. By signing the contract, Amelia agreed to serve without pay and to take no money, either from royalties or advertising contracts involving the flight. All income earned from these sources would help pay for the *Friendship*'s expenses. Mrs. Guest had paid all the expenses of the *Friendship* out of her own pocket. She hoped that the money earned from various promotions would help reimburse her. As captain of the flight, Amelia's word was to be final on all decisions once the plane was airborne. Wilmer L. "Bill" Stultz was named pilot of the *Friendship*, with Louis Gower as alternate pilot and Louis "Slim" Gordon as mechanic. Amelia was told she would share flying duties with Stultz or Gower.

The possibility of failure was not lost on Amelia. She knew it was possible — maybe even probable — that she might perish on the flight, as others had before her. Yet she had made up her mind; it would be worth the risk. She began to write farewell notes to every member of her immediate family, in case she didn't live through the flight.

George Putnam and the rest of the people associated with the *Friendship* warned Amelia to keep silent about the impending flight. Several others had already attempted the route from America to England. Putnam and his committee were afraid even more pilots would attempt the flight before the *Friendship* if the plans were made public. Amelia was not even permitted to examine the Fokker airplane that would carry her across the ocean, even though it was being serviced in east Boston, where she spent many Sunday afternoons.

Sam Chapman, who remained a close friend despite the fact that he and Amelia didn't get married, was the only person close to Amelia who knew what she planned to do. She said nothing to her family. Sam was to inform them of the attempt once she left Boston. Amelia's feelings about her adventure were stated clearly in the notes she wrote to them. Should she perish on the flight, these notes would remind each of them of her love, and of her absolute belief that it was important for her to have attempted the crossing.

At the same time Amelia was preparing for the flight to England, another young woman named Mabel Boll was planning to cross the Atlantic Ocean by plane. Mabel was an experienced horsewoman, though not a pilot. She was also a flamboyant figure who loved capturing news headlines for herself. Perhaps that's why she was sometimes called the "Diamond Queen." In any case, Mabel Boll announced boldly that *she* was going to be the first woman to cross the Atlantic, with none other than Bill Stultz as her pilot!

Mabel's announcement caused a big stir among the crew that was preparing the *Friendship* for Amelia's big flight. Would the Diamond Queen beat Amelia across the Atlantic? And how could Bill Stultz fly both missions? Stultz had to choose between flying with Amelia or with Mabel. When Stultz consulted with Commander Byrd about what to do, Byrd suggested that Stultz keep his word to Amelia. Stultz agreed, and Mabel Boll was forced to postpone her flight until she could find another pilot.

Amelia soon had another problem to resolve. Stultz, in addition to having an excellent flying record, also had a reputation as an alcoholic. When he became frustrated with his problems, he drowned them in liquor. Sometimes he disappeared on drinking sprees. Amelia's experience with her father helped her cope with Stultz during such times. She made it her responsibility to find him when he disappeared, and to sober him up if necessary.

Finally, in the late spring of 1928, the aircraft was ready for takeoff. A large, orange-colored plane with three motors, the craft had originally been the property of Commander Byrd before being sold to Mrs. Guest. Mechanics had checked it out thoroughly, and they believed it was hardy enough to endure an ocean crossing.

Next, the weather turned uncooperative. There were days of fog and heavy clouds, the kind of weather that could seriously hinder the chances of reaching England safely. If Boston's weather was clear, the weather in Newfoundland, their first stop, was cloudy. If Newfoundland was clear, Boston was fogged in. The entire group postponed flight after flight and waited.

Amelia grew so frustrated that she returned to work at Denison House. At least it would give her something to do. She spent some of this time making final preparations for the possibility of her death. She rented a safety-deposit box in a bank in West Medford and placed her will in it. If she failed to return, her mother would receive the balance of her small estate once all debts were paid. The notes she had written to her mother, father, and sister were also stored in the box. All were cheerful and optimistic, such as this one, which she wrote to her father:

Dear Dad:

Hooray for the last grand adventure! I wish I had won, but it was worth while anyway. You know that.

I have no faith that we'll meet anywhere again, but I wish we might.

Anyway, good-by and good luck to you.

Affectionately, yr doctor,

Meel

Finally, on June 3, the weather reports gave the crew reason to hope again. In the early-morning light of June 4, Amelia, in her dark-brown breeches, a white silk blouse, and high-laced boats, headed for Boston Harbor. The orange-colored *Friendship*, pontoons attached to its wheels, gleamed as it floated. The pontoons would allow the plane to float in the ocean, if that became necessary.

Amelia, Bill Stultz, Lou Gower, and Slim Gordon climbed on the tugboat *Sadie Rose*, which chugged them out to the plane. Bill Stultz started the engine, but he could not raise the plane from the water. The *Friendship* was simply too heavy to take off. The crew jettisoned a five-gallon tank of gasoline, but still Stultz was unable to lift the plane out. Then Lou Gower, who weighed 168 pounds, got out. His weight made the difference. When Stultz gunned the engines, the plane rose into the air. They were on their way to Trepassey, Newfoundland.

Meanwhile, the great secret was no longer a secret. As soon as the *Friendship* lifted out of sight, the

Boston newspapers began their coverage of the historic flight. In fact, that's how Amy and Muriel Earhart learned about the project! Sam had promised Amelia to break the news to them as soon as the *Friendship* departed, but the reporters got to the women first.

On June 5, the *Friendship* landed safely at Halifax, Newfoundland, in heavy fog. From there, Amelia sent her mother a cable: *Know you will understand why I could not tell plans of flight. Don't worry. No matter what happens it will have been worth the trying. Love, A.* Her mother's answer must have lifted Amelia's heart: *We are not worrying. Wish I were with you. Good luck and cheerio. Love, Mother.*

The plan called for the fliers to take off over the Atlantic Ocean the next day. Once again, however, bad weather forced a delay. Days passed without a break in the fog and the cold. The delay was not anyone's fault, but the media was not kind to Amelia and her crew. Some newspapers lost interest in the venture, while others criticized Amelia and the other crew members unmercifully. Bill Stultz began to drown his sorrows with a bottle of whiskey.

Amelia tried to ignore the critical stories she read in the papers. She received some comfort from George Putnam, who sent her a comforting telegram that ended: *Suggest you turn in and have your laundering done.* Amelia's answer, in her usual upbeat fashion, was: *Thanks fatherly telegram. No washing necessary. Socks underwear worn out. Shirt lost to Slim at rummy. Cheerio, AE.*

While they waited for the weather to clear, the crew

turned its attention to discarding every possible item that could weigh down the plane during its crossing. Amelia herself carried only a toothbrush, a small packet of food, notes from Americans to friends in England, and Commander Byrd's book, *Skyward*. The book was inscribed to Mrs. Guest and would be presented to her on their arrival. Everything that could be thrown out was left behind, including blankets, books, and food.

The weather finally broke two weeks later. Amelia and the crew took quick advantage of clear skies. On June 17, 1928, at eleven in the morning, the *Friendship* took wing for England.

Amelia did not do any of the flying, as she had hoped and expected to do when she first accepted the challenge. Instead, her job for the flight was to keep the plane's log and to watch the vast, almost indescribable beauties of the world outside her window. She was painfully disappointed that she would not be permitted to take the controls, for doing the flying was as important to her as being in the plane itself. She felt that the true glory of the adventure lay in acting as pilot.

The hours of the flight tolled on. After a while, Stultz turned over the controls to Slim. While Stultz napped, Amelia zipped herself into a fur-lined, full-length flight suit that had been lent to her for the occasion. Commander Byrd had advised her to use it, knowing that the air was usually very cold at the high altitudes at which planes flew. Indeed, Amelia had started to feel chilly. After she had snuggled into the suit, she felt completely comfortable once more.

By now the plane was flying on emergency fuel. They had at best only an hour or so left before they would run out of fuel completely. Yet they could not see land. Tension crept over everyone in the cockpit. Rain and fog rolled over them, and the crew began to wonder whether they would find a safe landing place at all. Where was England? Suddenly, all three spotted a curve of land dotted with factory chimneys. They had found land! They were going to make it!

Stultz maneuvered the *Friendship* carefully into a small bay with smooth water. As he did so, Amelia wrote in the log: *20 hours, 40 minutes out of Trepassey* Friendship *down safely in harbor of* ________________. She had no idea what harbor it was, but it didn't matter. The *Friendship* had come home.

For an hour, the plane bobbed on the waves out at sea. Rain poured in sheets around them, until a small boat reached them and brought the weary fliers ashore. Amelia, Bill Stultz, and Slim Gordon discovered they were in Burry Port, Wales, far from Southampton, England, where they had intended to land. They sent word to Captain Railey, who was waiting for them in Southampton. He immediately took a launch to Burry Port.

Pandemonium broke out as the three adventurers stepped ashore. Two hundred people tore at the fliers' clothes on the wharf. The spectators rushed forward to touch the fliers, talk to them, get their autographs. Everywhere she looked, Amelia was surrounded by beaming, admiring faces. She was the true heroine of the voyage. After all, men had already crossed the

Atlantic, but never a woman! Men, women, and children cried out to her, begging for her attention, for a smile, or for a word. She had not flown as much as a minute of the flight herself. But Amelia Earhart was suddenly the most famous woman aviator in the world.

6

Lady Lindy

CLEARLY, AMELIA'S LIFE would never be the same again. The social worker who had walked the streets of Boston with immigrant children clinging to her hands would soon have the most recognized face in the world.

After a night's rest at Burry Port, the crew of the *Friendship* climbed once more into the plane, along with Captain Railey, and took off for Southampton, England, their original destination. On their arrival, a launch carrying Mrs. Frederick Guest, Mrs. Guest's son, and Hubert Scott Payne of Imperial Airways met the plane. Mrs. Guest and the others officially greeted the fliers. It was Amelia's first meeting with the woman who had changed her life. Mrs. Guest invited Amelia to stay with her at her home on London's fashionable Park Lane. Amelia liked Mrs. Guest right away, and she accepted gratefully.

The public, including U. S. President Calvin Coolidge, believed that Amelia had been instrumental in guiding the *Friendship* to safety. This made Amelia feel somewhat awkward. She insisted over and over to reporters, friends, and everyone who asked that she had nothing to do with the success of the flight. When President and Mrs. Coolidge sent a telegram congratulating her, Amelia replied, "Success entirely due great skill of Mr. Stultz." She told her many new fans that she had been nothing more than a "sack of potatoes" while the *Friendship* was aloft.

Yet it was Amelia who attracted crowds of admirers, not Stultz or Gordon. English women's clubs begged her to appear at their luncheons. She met and danced with the Prince of Wales, heir to the throne of England. She could not shop in stores without being approached for autographs and photos.

Part of the fame was very pleasant to Amelia. One day she went shopping at Selfridge's, a department store in London, intending to choose articles to wear in London and on the ship back to the States. (She had not brought much with her on the plane.) The owner of the store, Gordon Selfridge, greeted Amelia personally and refused to allow her to pay for anything. Mrs. Guest made her a present of a beautiful leather handbag. Lady Astor, an American-born member of the British Parliament, spent several hours with her. Wherever she turned, Amelia had dinner, luncheon, and party invitations thrust upon her.

From her point of view, perhaps Amelia's greatest opportunity in London was meeting Lady Mary Heath, England's most famous woman pilot. Lady

Mary owned an Avro Avian Moth airplane she had recently flown from Capetown, South Africa, to London. Like Amelia, Lady Mary was a courageous woman pioneer whose little open plane had flown through many miles of "uncharted" airspace, or airspace that had not yet been explored. Her journey, though shorter and less spectacular than Amelia's, was no less dangerous. Like Amelia, Mary had been welcomed in London as a heroine when she stepped from her Avro at the end of her flight.

Amelia and Mary took the Avro up for a short flight. While they were aloft, Mary mentioned that she was interested in selling the plane. Amelia asked whether Mary would allow her to buy it. Though she didn't have enough money yet, Amelia offered to pay for it as soon as she had the funds. Mary agreed. When Amelia set sail for New York on the S.S. *President Roosevelt*, Mary Heath's Avro set sail with her.

When the *President Roosevelt* docked in New York, pandemonium erupted there, too. Amelia thought her English reception had been exciting, but it was nothing compared to the screams and cheers of the New Yorkers who greeted her. Like so many heroes, Amelia, along with Bill Stultz and Slim Gordon, was given a ticker-tape parade through the streets. Invitations poured in, far too many for the three fliers to accept. They limited their activities to events in Boston, Medford, and Chicago.

Amelia, in particular, was embarrassed by the adulation that poured down on her from all sides. While she appreciated the kindness and sincerity of the people who approached her, all the attention made her increasingly unhappy. She felt like a fraud. Still, no matter how often she persisted in saying that she had not actually piloted the *Friendship*, no one believed that she had not been vital to the success of the Atlantic crossing.

Amelia was too honest with herself not to feel ashamed of her fame. After all, she had become a celebrity for something she hadn't even done. She hoped that someday she would be able to earn the fame she deserved on her own merit as a pilot. "The next time I fly anywhere, I shall do it alone!" she told close friends with determination.

Meanwhile, Amelia had contracts to honor. She had agreed to write a book about the experiences of the crew on the *Friendship* flight. George Putnam, whom she called G.P., had arranged the contract and offered her the use of his home in Rye, New York, while she wrote it. His wife, Dorothy, and their two children provided a homelike atmosphere for Amelia. G.P. also volunteered the use of his secretary, who helped Amelia make sense of her notes and organize the book.

It was G.P. who christened Amelia "Lady Lindy," a label that would stick until the end of her life. Because of her daring and courage, and also because Amelia had a more-than-passing physical resemblance to Charles Lindbergh, G.P. thought the nickname was appropriate. The press picked up the name immediately, and soon the tag was so famous that it was even celebrated in a song.

Amelia had expected to take only a two-week leave from Denison House. But too much was happening in

The Bettmann Archive

After her successful transatlantic flight in 1928, Amelia Earhart rode in triumph through New York City. It was popularly said that Amelia had both the courage and the looks of Charles Lindbergh, who in 1926 became the first man to fly solo across the Atlantic Ocean. Amelia was hailed in the press as "Lady Lindy."

her life for that to be possible. She began to realize it might take as long as six months before she could return to her duties there. It even occurred to her that she might never return. Opportunities were falling in Amelia's lap, and she felt it was important to take advantage of them. She wanted to do all she could to advance the cause of women in aviation. The more she accomplished in the field, the greater value she felt she would be to society in general, and to herself in particular.

Amelia completed her book, *20 Hours, 40 Minutes*, and then turned her attention to lecturing. G.P. scheduled her for an exhausting lecture-tour in the winter of 1928-1929. She spoke enthusiastically at clubs, professional organizations, and large business gatherings about the importance of aviation and its growing influence on the world. She also accepted an invitation from *McCall's* magazine to act as its aviation editor. Amelia believed that access to the public through a popular monthly women's magazine would greatly increase women's interest in flying.

A certain amount of controversy grew out of some of Amelia's publicity. A cigarette company promoted an ad stating that Amelia and the crew of the *Friendship* smoked their cigarettes. In fact, Amelia did not smoke, and asked to be left out of the ad. The company refused, saying that the ad was of little value without her, the most famous crew member. Amelia knew that Bill Stultz and Slim Gordon would profit handsomely from the ad, and agreed to endorse it for their sake. Her own share of the money was earmarked for Commander Byrd's second expedition to the Antarctic.

The ad sparked controversy for a second reason. In 1929, the idea of women smoking, especially in public and in advertisements, was daring and bold. *McCall's* disapproved of portraying Amelia as a smoker so strongly that they canceled their contract with her! G.P. protested the magazine's treatment of Amelia, and persuaded the editors to reconsider. At the same time, however, *Cosmopolitan* magazine, a rival of *McCalls*, offered Amelia a position as an associate editor. Amelia accepted. As part of her job, she would write articles about flying and keep an office in their building.

The articles Amelia wrote for *Cosmopolitan* were lively, informative, and fun to read. From the beginning, the public loved them. In her writing, Amelia praised the courage of pioneers in the skies and hailed the joys of flying. She also expressed her opinions on controversies that arose in the field of aviation. At one point, female pilot Helen Richey was barred from flying for Central Airways because she could not join the pilots' union, which admitted only men. Amelia wrote a provocative article pointing out that Miss Richey was as qualified a flier as any man in the union, and that therefore she should be permitted to join the union and seek employment with the airlines. However, Richey was not admitted to the union.

Amelia did not have enough time to resume her duties at Denison House, but she could not bring herself to abandon the people there. Her former supervisor, Marion Perkins, suggested that Amelia join the Denison House board of directors. That way, she could remain as active as possible in the project without having to appear at every meeting. Amelia agreed happily, and remained on the board for years.

Amelia also accepted opportunities to model and design sports clothes of her own, and she agreed to endorse a variety of products, from chewing gum to luggage. To this day, Amelia Earhart luggage, manufactured by the Baltimore Luggage Company, still attracts many buyers each year.

One of Amelia's endorsements involved more than simply smiling at a camera. The Beech-Nut Packing Company asked Amelia to demonstrate its new autogiro, a heavier-than-air craft which was the forerunner of today's helicopter. In late June of 1931, Amelia set a new altitude record for autogiros of 15,000 feet. That same day, she broke her own record, flying the autogiro at 18,415 feet. Once she was satisfied with the altitude record, she took off again in the autogiro and flew to Los Angeles. That trip established Amelia as the first autogiro transcontinental pilot and the first woman to cross the United States in that type of craft.

There was no question that Amelia Earhart had become a very famous woman, a heroine for people all around the world. One of the advantages of her fame was the opportunity it provided for Amelia to do good deeds for others. Without drawing any attention to herself, Amelia quietly offered help to a number of people. She paid the rest of the mortgage due on her father's house so he could live worry-free with his second wife, Helen. She sent clothes to a little girl in Chicago whose family always dressed her in hand-me-downs. She paid for treatment for an alcoholic friend of hers. None of these "lifts," as Amelia called them, ever appeared in print. She did them simply because she delighted in helping others.

Meanwhile, Amelia's personal life was moving toward a climax. Sam Chapman had dropped out of her life after her overnight fame. Now, another man was pursuing her. This time, she would finally say yes.

7

The Perfect Team

GEORGE PALMER PUTNAM II fascinated Amelia from the first time she met him at her all-important interview with the men who had done so much to organize and publicize the *Friendship*'s Atlantic crossing. Ten years older than Amelia, G.P. was the grandson of one of the publishing world's most prominent men, George Putnam, Sr., who had founded G.P. Putnam Sons, Inc., and the son of George Haven Putnam. The younger G.P. now ran the publishing company himself. Under his guiding hand, the company had signed many prominent authors and grown tremendously. He had a particular preference for authors who wrote adventure books. Adventure, in books *and* in real life, fascinated G.P.

It captivated him so much that he did some exploring of his own. In 1925, the American Museum of Natural History asked him to explore Greenland, and

G.P. was happy to oblige. At the end of the expedition G.P. became an author, helping his son David write *David Goes to Greenland*, a children's book about his adventure in exploring. G.P. would go on to write ten books, including a biography of Amelia.

G.P.'s special talents, however, were in the areas of promotion and publicity. In particular, G.P. had a genius for attracting reporters with a new scheme or raising funds for a risky new venture. He was the perfect man to turn to when Mrs. Guest needed to organize a flight across the Atlantic. After the crossing had been successfully completed, G.P. became the perfect man to boost Amelia's career.

Amelia and G.P. had many common interests. Aside from their mutual delight in high adventure, they both enjoyed art, music, and the theater. They possessed the same boundless energy. G.P.'s constant enthusiasm and ingenuity were great assets. He could organize a lecture schedule for Amelia or raise funds for some farfetched plan she wanted to try in order to further the cause of women in aviation. No matter how crazy her wishes seemed, G.P. always found a practical way to manage them.

When he first met Amelia in 1928, G.P. was married to a New York socialite named Dorothy Binney. The couple had two sons. In fact, Amelia dedicated her book, *20 Hours, 40 Minutes*, to Dorothy with thanks for her hospitality during its writing. In 1930, however, Dorothy and the two boys left G.P.'s home in Rye, New York, to move to Florida. Soon after, the Putnams were divorced.

G.P. began to woo Amelia. He continued to orchestrate her public life. He set up contracts and interviews, helped her focus on her goals, and taught her how to achieve them. In order to glamorize her appearance he instructed Amelia to smile with her lips closed. This would conceal the gap in her front teeth. He set up special photo sessions with prestigious magazines such as *Vanity Fair*, for whom Amelia modeled chic and sporty clothing. He was always there to lend a hand or offer comfort when she needed it. G.P. became a good friend who was invaluable to her.

One of his most memorable gestures was helping Amelia to enter the first Women's Air Derby in 1929. This special event was a race from Santa Monica, California, to Cleveland, Ohio. Because only women pilots could enter, the race was immediately nicknamed "The Powder Puff Derby." It was women's first organized recognition in aviation, and Amelia eagerly entered the race. She finished third, behind Louise Thaden of Pittsburgh and Gladys O'Donnell of Long Beach, California.

Persistence may have been G.P.'s greatest talent. By his own admission, he proposed to Amelia at least six times between 1930 and 1931. She refused five times.

Amelia, who was now in her early thirties, was not sure marriage was such a good thing. From what she had seen, women had great need for security in marriage. This troubled Amelia. She felt that her mother's need to "settle down" with an assured income had contributed to her father's inability to pursue his goal of becoming a Supreme Court justice.

Sam had wanted Amelia to give up flying. Amelia believed that, in the long run, asking for such security destroyed both men and women. She felt it forced everyone involved in a marriage to give up on their dreams. If that was what being married meant, then Amelia wanted no part of it for herself. She wanted to fly and she wanted to encourage women to enter the field of aviation. These goals did not seem compatible with marriage.

G.P., however, understood Amelia's reluctance to settle down. He assured her that if they married, she *would* continue to fly, as often as she pleased. He would do everything possible to help her. He would continue to act as her promoter and manager for as long as she wanted to continue being a pilot. Month after month, G.P. worked to lessen Amelia's fears about marriage. Instead of trying to limit her world, he broadened it, by introducing her to the many different, well-known people he had met. He charmed her with his attention, and he refused to give up. G.P. made himself indispensable to Amelia in every way that was important to her.

When her father died of stomach cancer in 1930, Amelia turned to G.P. for solace. This tall, slim man, whom she thought was so self-confident and charming, was the man her father would have been had Edwin lived his life differently. That realization may have been the deciding factor in Amelia's change of heart.

By this time, Amelia had acquired a bright-red Lockheed Vega, a small high-winged plane with an open cockpit that seated only two. One morning,

UPI/Bettmann Newsphotos

Amelia with her husband, George Palmer Putnam, grandson of the founder of G.P. Putnam's Sons , at the opening of a new theater in New York City in 1932. Putnam, who ran the publishing house, thoroughly supported his wife's adventures, both financially and emotionally, and was a leading force behind arrangements for her flights and their publicity.

while she was in the hangar waiting for the plane to warm up, forty-three-year-old G.P. proposed again. This time, the thirty-three-year-old Amelia nodded yes, patted his arm quickly, and climbed into the cockpit.

When Amelia's mother learned of the engagement, she advised Amelia not to marry G.P. She disapproved of her daughter marrying a man who was ten years older than she was, and who was divorced. Once Amelia made up her mind, however, she intended to go through with it.

On the morning of February 7, 1931, she and G.P. were at his mother's house in Noank, Connecticut. It was their wedding day. Everything was ready except one thing. Amelia wanted to be certain that the terms of their marriage were spelled out clearly. She handed G.P. an extraordinary note:

> "In our life together, I shall not hold you to any medieval code of faithfulness to me, nor shall I consider myself bound to you similarly. If we can be honest I think the difficulties which arise may best be avoided.
>
> "Please let us not interfere with the other's work or play, nor let the world see our private joys or disagreements. In this connection I may have to keep some place where I can go to be myself now and then, for I cannot guarantee to endure at all times the confinements of even an attractive cage.
>
> "I must exact a cruel promise, and that is you will let me go in a year if we find no happiness together."

G.P. was perhaps startled at the terms Amelia had set, but he had no choice if he wanted to marry her. He nodded agreement, and the couple was married in a very private ceremony. Neither Muriel nor Amy was invited to the wedding.

The marriage was a success from the start. Though the couple sometimes appeared mismatched, they answered each other's needs perfectly. G.P. could be a hardheaded businessman, concerned only with making a profit, while Amelia, whom he called "A.E.," had a tendency to be less practical about money. Where she was softhearted and anxious to achieve goals that seemed impossible, he was organized and completely down-to-earth. While she flew, he supervised the details.

G.P. nurtured her career so insistently that Amelia became the best-known woman in the country. She insisted on keeping the name "Earhart" for her professional activities. Yet her private life as "Mrs. Putnam" gave her great satisfaction. She enjoyed the company of G.P.'s two sons and was genuinely affectionate with and interested in them. Both boys, in turn, adored her. She loved G.P.'s home in Rye. When their first year together was over, she did not ask for release from the marriage.

A few years after their wedding, a devastating fire broke out in the Rye house. Though the family was away and was thus spared any physical harm, many of Amelia's valuable mementos and much Putnam family history were lost in the blaze. Following the tragedy, G.P. and Amelia decided to move to the West Coast, where the clear weather would allow Amelia to fly far more often than she did on the stormy East Coast. They purchased a home in North Hollywood, California. In 1931, G.P. had sold his interest in the publishing company to his cousin Palmer, and accepted a position as head of the editorial board of a motion-picture company, Paramount Pictures. This gave him the freedom and flexibility to focus his enormous energies on Amelia's fame. He worked hard to enhance her reputation and make it grow.

G.P. loved having the limelight for himself, too. He adored having his picture taken and chatting with reporters. Newsmen always preferred Amelia to her husband, but they found to their sorrow that in order to get a few words from Amelia, they had to endure a few thousand from G.P.

Husband and wife clashed more on commercial matters than anything else. G.P. once arranged for a manufacturer to make "Amelia Earhart *Friendship* Flight Hats." These were to be replicas of the small tan hat that Amelia had worn on her return from the *Friendship*'s Atlantic crossing. G.P. decided that the hat would be trimmed with a dark-brown ribbon bearing Amelia's signature. It would cost three dollars, although it was worth little more than twenty-five cents. G.P. figured that children would buy it anyway, in order to own a replica of Amelia's signature. Amelia was in California at the time that G.P. negotiated this deal in New York. When she found out what he had done, she flatly refused to enter into the promotion. She insisted he cancel the contract at once, saying, "Forget it, George. I won't be a party to cheating youngsters. Adults are supposed to know better, but not kids."

G.P. Putnam's importance to Amelia's life and flying career cannot be overestimated. He gave Amelia her first opportunity to fly a world-record flight. He encouraged and promoted her effectively, and he supported her desire to blaze trails for women aviators. In turn, Amelia found G.P. to be a man she could respect and rely on. Without him, Amelia would have been unable to raise the funds necessary for her record-breaking flights. His presence was as important to her career as her own. Surely, for America's favorite female flier, there could be no better partner.

8

Atlantic Solo

AMELIA'S DESIRE TO make a record flight on her own was always at the back of her mind. Since she herself had not done any of the flying during the *Friendship* flight across the Atlantic, Amelia felt that she had not truly earned her fame. She wanted to prove, if only to herself, that she deserved the celebrity that was heaped on her.

Amelia knew that a record-breaking solo flight required careful preparation. She also believed that she needed to log far more flying hours than she already had in order to be qualified. With her usual energy and determination, she began to get ready herself.

One morning in early 1932, at the breakfast table, before their move to North Hollywood, Amelia said casually, "G.P., would you mind if I fly alone across the Atlantic?"

G.P was not surprised at the question. He knew that Amelia felt it was important to prove to herself that she *was* a true aviation pioneer, not just a phony that the public had taken to heart. She also believed strongly in pushing the known boundaries of aviation. The sooner men and women knew every square inch of sky above the earth, the sooner flying would become a routine matter for the public. This could only be accomplished by flying through uncharted airspace.

G.P. assured Amelia that he would support her wish. Immediately they began to assemble everything Amelia would need for such a dangerous expedition. The flight would cost a great deal of money, and they both agreed that the best way to raise it would be to promote the flight and Amelia. G.P. set himself the task of finding backers who would help finance the project. He arranged for advertisements for various products using Amelia's name and picture, and promised future promotions when Amelia returned. Soon he had secured the needed funds.

Preparation for the flight was not extensive. Amelia intended to carry very little with her. Having already flown the same route, she did not need much in the way of navigational help. She was to fly in her red Lockheed Vega. Most of the preparations were similar to what had been done for the *Friendship* flight. Amelia did decide to make one change, however. She had disliked the pontoons on the *Friendship*, which had been added in case they needed to land in the ocean. She refused to have them on the Vega during her solo flight.

Once more, bad weather intervened to delay the

flight. On Thursday, May 19, 1932, both Amelia and the weather were finally ready.

Amelia dressed neatly in tan jodhpurs, a white silk blouse, a light windbreaker, and a blue-and-brown scarf tied at her throat. She carried two cans of tomato juice in a knapsack and tucked a toothbrush and a comb in her pocket, along with a twenty-dollar bill to pay for telegrams she would be sending to America once she reached England.

The Lockheed Vega waited in a hangar at Teterboro Airport in New Jersey. That afternoon, Amelia and her crew, pilot Bernt Balchen and mechanic Eddie Gorski, left for Saint John, New Brunswick. Balchen piloted the Vega while Amelia conserved her energy with a nap. The next morning, the three flew from Saint John to Harbour Grace, Newfoundland. Gorski checked the plane carefully and refilled the fuel tanks. Everything was ready for Amelia's solo flight.

On Friday, May 20, five years to the day after Charles Lindbergh set off for Paris, Amelia climbed into the cockpit of her Vega. She gunned the plane's single engine, and at 7:12 P.M. the Vega rose into the evening sky. As darkness enveloped the plane and the stars and moon appeared, trouble developed. Amelia noticed with dismay that one of her instruments, the altimeter, was no longer functioning. Pilots refer to the altimeter for information about their altitude above sea level. Not knowing how high a plane is flying can cause disaster for pilots. This was the first time in twelve years of flying that Amelia had faced this problem. Then a mechanical problem developed, causing heat in the engine. Amelia knew that if the plane was tossed around in storm conditions, there might well be a fire.

In fact, Amelia's route took her straight into a storm cloud. The rain and heavy wind tossed the Vega around while Amelia fought the controls to maintain her course. For an hour she worked at keeping the plane steady. Finally, she obeyed the pilot's rule of thumb regarding storms: "When you fly into a storm, climb out of it or fly under it."

Amelia had to decide whether to go above or below the storm. Since she did not have an altimeter to give her accurate information, she decided to climb above the cloud. She was afraid that if she flew under it, she might accidentally plunge into the ocean. However, when the plane climbed, ice formed on the wings. This is one of the most dangerous conditions a flier can face. Ice on the wings makes it very difficult to control the plane. The only way to melt ice is to fly at a lower altitude. Amelia pushed the plane down, wondering every minute whether she would slam into the icy water below.

The heat from her engine became intense. Glancing back, Amelia saw a ring of fire burning behind her. It was far too late to try to return to Harbour Grace. She knew she would not possibly be able to set the plane down in the dark, and she feared an explosion with so much gasoline aboard. Despite the possibility that she might misjudge the level of the sea, Amelia kept the Vega low, almost skimming the ocean waves. She preferred to drown rather than burn to death in mid-air.

In complete darkness, flying with the aid of instruments and unable to see anything around her, Amelia chose to ignore her feelings of dread. As she said later, "Probably, if I had been able to see what was happening on the outside during the night, I would have had heart failure then and there; but, as I could not see, I carried on."

As daylight dawned over the ocean, Amelia could at last see clearly enough to assess her situation. The Vega was flying between two layers of heavy clouds. She knew that the winds had battered her far off course. Fire still burned in her engine. She had no choice but to find a safe place to land. Then, moments later, she sighted land! Amelia was still lost, of course, but nonetheless she was thrilled. If she could land safely she would have accomplished her goal!

The rule for fliers who were lost in the United States was very clear and very logical: Find a railroad and follow the tracks, which would eventually lead to a town somewhere on the route. Amelia, however, was flying over Ireland, and the rule simply did not apply; railroad tracks there did not necessarily lead to a town. This didn't matter to Amelia. She was relieved to be flying over land at last, and ready to set down whenever she noticed a likely place.

Finally, she spotted a flat piece of pasture. Circling carefully over it, she landed among curious brown-and-white cows. It was May 21, 1932. Fourteen hours and fifty-six minutes had passed since her takeoff from Harbour Grace, Newfoundland. Amelia was thirty-four years old. She had achieved her mark of distinction on her own terms. No one, least of all

Amelia herself, could call her a "sack of potatoes" again.

James Gallagher, the farmer who owned the pasture where the Vega landed, near Londonderry, stared in surprise at the slender young woman who popped out of the cockpit of the red plane.

"Hi," she greeted him jauntily. "I've come from America!"

The farmer stared at her. "Do ye be tellin' me that now?"

The news that Amelia had successfully navigated the Atlantic Ocean alone sent crowds into a frenzy. When she arrived at the American Embassy in London, she was greeted with stacks of telegrams and phone messages. Colonel and Mrs. Charles Lindbergh sent their congratulations, along with friends and relatives from America. Amelia grinned at the cable from Phil Cooper, who owned the dry-cleaning store she and G.P. used in Rye. The cable read: *Knew you would do it. I never lose a customer.*

Again Amelia mingled with royalty, meeting privately with notables such as the Prince of Wales. Despite the careful scheduling of the prince's appointments, he ignored them to spend an extra half-hour with Amelia. The prince told her he envied her the freedom of flying. He said that he himself would love to own and pilot his own plane, but as heir to the throne, British tradition would not permit him such a risky luxury. He was also extremely pleased that Amelia's second flight, like her first, had ended in the United Kingdom.

Honors poured in from every country in the world.

Amelia received the French Cross of the Legion of Honor, a Certificate of Honorary Membership in the British Guild of Airpilots and Navigators, and the National Geographic Society's Special Gold Medal of achievement. Italy honored her with its Balbo medal. The United States Congress presented her with the Distinguished Flying Cross, the first woman to be so honored.

Still, amidst all the glory, Amelia herself felt compelled to point out, with her usual honesty, "My flight has added nothing to aviation. After all, literally hundreds have crossed the Atlantic by air . . . However, I hope that the flight has meant something to women in aviation. If it has, I shall feel it was justified, but I can't claim anything else."

By now, Amelia and G.P. had met President Herbert Hoover and the king and queen of Belgium, and were being received with enthusiasm all over the world. Yet Amelia remained the same woman she had always been, her feet on the ground and her eyes on the stars. There was still so much to be done! She appreciated the honors, but preferred to daydream about the next flight she would attempt. She could push women further if she continued to inspire them with her accomplishments. It was important to remain a spokesperson for women in the skies. Of course, Amelia would be the first to admit that her record flights had another motive that was just as important. She did what she did "for the fun of it!"

9

An Inspiration to Women

AMELIA ACCOMPLISHED AS much on the ground as she did in the air. Though her daring flights brought her into public focus, she managed a great deal of influence in women's organizations as well. Behind the public facade of daredevil aviator, Amelia maintained her interest in women's achievements and her desire to lend her name and talents to the cause of greater equality between the sexes.

One of Amelia's most important contributions grew out of her participation in the National Air Races and Aeronautical Expo in Cleveland, Ohio, in 1929. As Amelia and other female pilots stood in a hangar relaxing, they discussed women's role in aviation. They agreed that there should be an organization exclusively for women pilots. Such an organization, they felt, should promote "good fellowship, jobs, and a central office and files on women in aviation."

Amelia and several other female flyers decided to send out membership invitations to every one of the 117 licensed women pilots in the United States. Since Amelia had a secretary, she volunteered her services to send out the invitations. The women agreed to meet with as many others as were willing to join.

Names for the young organization were suggested. Some of the suggestions were rather flamboyant, such as The Gadflies or The Noisy Birdwomen. Amelia made the suggestion that the group finally adopted. She said they should count the number of women who actually joined as charter members, and use that number as their name. When ninety-nine pilots joined the new organization, it became known as the Ninety-Nines (also referred to as the 99s).

The group met for the first time in November, 1929, at Curtiss Field in Long Island, New York, and elected Amelia their first president. Her ideals became those of the Ninety-Nines: recognition for women pilots, encouragement to women everywhere to take to the skies, equal opportunities and equal pay for women doing the same work men did in aviation. The Ninety-Nines sponsored air races and forums for its members, and took an active role in providing financial aid to members who were unable to advance without help. In 1941, the Ninety-Nines established an Amelia Earhart Memorial Scholarship Fund, with the money earmarked for women pilots who wanted to advance their education or do research in any area of aviation.

The Ninety-Nines are still in existence today. They are a thriving community of almost 7,000 members, and there are chapters of the association in countries all over the world. Although all the original members of the Ninety-Nines were vocal and opinionated, Amelia held a special place among them as their most famous member. Often, when they disagreed, Amelia kept the Ninety-Nines together with her own quiet reason, dedication, and personal strength. At the Oklahoma City archives, there is an entire section devoted completely to Amelia and her achievements.

Amelia's dedication to women in aviation went even further. She became involved in other associations, among them Zonta International. Zonta International was an organization for professional women based on a high code of both professional and personal ethics. To be invited to join was an honor. After the *Friendship* flight in 1928, Amelia received an invitation to join the Boston chapter of the organization. She was attracted to Zonta's ideals and its international membership, and she readily joined its ranks. Zonta was a nonflying organization — the only nonflying club to which Amelia belonged — but she chose membership in the category of aviation. In 1930, she became an active member of the larger New York chapter. In 1932, she won the Zonta Club trophy in recognition of her Atlantic solo flight. In 1938, Zonta established the Amelia Earhart Graduate Scholarship for Women. Through the fund, Zonta gave "lifts" in Amelia's name that helped women in practical ways.

Another practical application of Amelia's flying expertise was her decision in 1928 to join the board of directors of Dennison Aircraft, which was a brand-new small commercial airline. Amelia was the first woman ever to sit as a board member of any commercial airline. She was listed as one of the five incorporators of the tiny company.

In 1935, Amelia participated in a speaking engagement that led to a whole new phase in her career. She and a group of other speakers had been invited by the New York *Herald Tribune* to address the subject, "Women and the Changing World."

The world had changed dramatically following World War I. Many young people who grew to maturity during the war had been deeply scarred by the conflict. They refused to look into the future because they believed it would be as horrible as the war had been. They were disillusioned and rootless, they lived for the moment, and they became known as the "Lost Generation." Instead of working at steady jobs, they became involved in more unconventional pastimes, such as playing jazz or marching in peace demonstrations.

At thirty-seven, Amelia was older than most of these young people. She did not understand their lack of passion or their boredom. But she defended their right to enjoy jazz, modern art, and peace demonstrations. It was for that reason that she agreed to speak on the panel.

One member of the audience that night was Dr. Edward C. Elliott, the president of Purdue University in Lafayette, Indiana. Dr. Elliott was impressed with Amelia. He believed that her understanding of young people made her a natural choice to serve on his campus. Specifically, he felt that the 800 or so female students at Purdue could benefit from the guidance of a woman like Amelia. Would she consider a position there?

Amelia accepted Dr. Elliott's offer. On September 1, 1935, she joined the Purdue faculty as an adviser in careers for women and a consultant in the aeronautics & aviation department. Her original idea was to spend one week each month on campus. But Amelia jumped into her new job with her typical enthusiasm and dedication. She soon took on more work at the university, and began to spend more and more of her time there.

Amelia loved her work at Purdue. It gave her an opportunity to meet young female students and hear their opinions. It also gave her the chance to encourage them in careers that until then had been reserved exclusively for men. About twenty young women told Amelia that they wanted to learn to fly. Amelia was delighted to hear this, and she urged them to pursue their dream. She warned them that it would be difficult to earn a living in aviation. Nonetheless, flying for the fun of it was a worthwhile goal.

The Purdue trustees showed their appreciation of all that Amelia had done for their school by setting up a trust fund for aeronautical research and setting a goal of $50,000 for the fund's endowment. It was such a success that within a year, they had raised the full amount. The Purdue Research Foundation, which set up the Amelia Earhart Fund, would later donate over $40,000 to buy Amelia a new plane.

As the years passed, Amelia continued to write and publish articles and books. Her second book, *For the*

Fun of It, written in 1932, described her solo Atlantic flight. She also penned numerous articles for *Cosmopolitan* magazine, where she remained aviation editor for three years. Her poem "Courage," written while she was a social worker at Denison House and first published in 1928, had become famous. In it she expressed her feelings about the importance of taking risks in life:

> Courage is the price that life exacts for granting peace.
> The soul that knows it not knows no release
> From little things:
> Knows not the livid loneliness of fear
> Nor mountain heights, where bitter joy can hear
> The sounds of wings.
> How can life grant us boon of living, compensate
> For dull gray ugliness and pregnant hate
> Unless we dare
> The soul's dominion? Each time we make a choice, we pay
> With courage to behold resistless day
> And count it fair.

Amelia's articles and books reached people all over the world, and helped to advance aviation and the cause of women. In addition, the actual profits from these endeavors provided funds she could use for future aviation projects. After all, Amelia had not yet accomplished all she wanted to. As always, there was another goal worth reaching out for.

10

The Endless Horizon

THE ATLANTIC SOLO was just the beginning of a string of amazing achievements by Amelia Earhart. In the five years following that flight, she made a series of daring and dangerous flights that broadened the scope of aviation far beyond its earlier boundaries. "Lady Lindy" was more than earning her nickname now.

After her Atlantic solo flight, Amelia was delighted to learn that Eleanor Roosevelt, the wife of President Franklin D. Roosevelt, was interested in flying. After dinner at the White House one evening, Amelia offered to take Mrs. Roosevelt on her first flight, right then and there. The First Lady agreed. Amelia was wearing a white evening dress and formal white gloves when she took Mrs. Roosevelt up in her plane; Mrs. Roosevelt was similarly attired. They flew from Washington, D.C., toward Baltimore, Maryland, and then circled back to the White House.

The First Lady loved her flight and promptly announced that she wanted to learn to fly herself. Naturally, Amelia was delighted. Here was one of the most famous women in the United States announcing that she thought flying was fun! It was wonderful publicity for the cause of women in aviation. This was exactly what Amelia had been trying to achieve for years.

Mrs. Roosevelt actually did get her student pilot's permit, and Amelia promised to teach the First Lady herself. But as it turned out, Mrs. Roosevelt could not continue with instruction. She was, after all, the President's wife. Flying was too risky for a woman in such an important position. However, Amelia and Mrs. Roosevelt became good friends. The First Lady supported Amelia in all of her flying ventures from that time on.

In 1934, Amelia was given the chance to achieve another first in aviation. That year, sugar-growers and businesspeople in Hawaii offered a $10,000 prize to any pilot who would fly solo from Hawaii to the American mainland. Hawaii was isolated from the rest of the United States by thousands of miles of ocean. The flight was to be a means of publicizing the Hawaiian islands, in the hopes of drawing both trade and tourism.

Such a flight was considered dangerous and difficult. Any pilot attempting to fly over such a vast expanse of water in the small planes available at that time was risking disaster. In fact, no flier had yet crossed the Pacific successfully. The flight required many hours in the air without refueling and with few landmarks a pilot could use to check his bearings. It was a challenge to even the most courageous fliers. It was the kind of challenge Amelia lived for. She stepped forward to meet it.

Paul Mantz, a veteran pilot who set up and flew airplane stunts for Hollywood movies, acted as Amelia's technical adviser. For six months, he and Amelia made plans and navigational decisions and did mechanical work on Amelia's Lockheed Vega. Finally, in December 1934, Amelia and Paul boarded a ship for Honolulu, with the Vega lashed to the deck. G.P., Mantz's wife Myrtle, and Amelia's mechanic, Ernie Tissot, accompanied them.

Paul Mantz had prepared Amelia well. He had installed a complete life-preservation system in the plane, including a rubber raft, hatchet, and sheath knife Amelia could use to cut her way to the raft, which carried emergency rations and distress-signal equipment. Should the Vega come down in the ocean, Amelia would have food and survival equipment that would last until she could be rescued.

In Hawaii, Paul and Ernie went over the plane for two weeks before they pronounced it ready. However, on Amelia's projected takeoff date, January 11, 1935, rain poured down for most of the day. The Vega was carrying a heavy fuel load — 500 gallons. Since there were no concrete runways at all in Hawaii, the Vega would be rolling on wet grass in order to take off, a difficult maneuver.

When Amelia tested the Vega's engine in the late afternoon, it sounded fine. She decided she *could* lift the plane off the ground, wet grass or not. At around 4:30 P.M., she ran up the motor and gunned the plane down the runway. Once more, she was on her way.

Like her Atlantic flights, Amelia's flight from Hawaii to the American mainland was flown almost entirely over an ocean. Amelia did not expect to see ships below her, and she did not believe the ships would be able to spot her, either. Yet two ships did, and Amelia turned on her landing lights to wink at them.

The rest of her communication came through her radio set. She was able to pick up commercial radio stations, which knew about her flight. When she was not broadcasting, Amelia listened to classical music. Station KGU in Honolulu interrupted its program to put G.P. on the air, and he spoke to his wife as though she was in the next room.

Not all the communication was so perfectly clear. One of the stations on the American mainland heard Amelia say, "I am getting tired of this fog." But the radio announcers picked up only part of the sentence, "I am tired." Many worried listeners imagined touching pictures of a spirited young woman in the Pacific battling exhaustion to reach San Francisco safely.

Finally, Amelia spotted a ship, the *President Pierce*, which she was told was a few hundred miles from San Francisco. An hour later, she spotted Pillar Point, which was less than fifty miles south of the Golden Gate Bridge in San Francisco. The Vega soared over San Francisco Bay to the Oakland Airport, where Amelia was greeted with screams and cries of joy. Seventeen hours and seven minutes after setting off from Honolulu, she had done it again. She was the first person to fly from Hawaii to California, the first person to solo anywhere in the Pacific Ocean, and the first person to have soloed over both the Atlantic and the Pacific Oceans.

Amelia added to her achievements with a goodwill flight in April 1935. A month earlier, she had met the General Consul of Mexico, Eduardo Vellasenor, in New York City. Vellasenor suggested that it would be a wonderful gesture if Amelia made the first flight by a woman aviator between Mexico and the United States. She agreed, and G.P. went to Mexico City to arrange it.

On April 20, 1935, Amelia's bright red Vega lifted off from Burbank Airport in southern California. Thirteen hours and twenty-three minutes later, the Vega set down at Valbuena Airport in Mexico City. Again she met with wild acclaim. She was the first person to solo from Los Angeles to Mexico City.

Amelia and G.P. vacationed in Mexico for a few days, meeting with the Mexican president, La'zaro Ca'rdenas, and chatting with some Mexican women about their life ambitions. The Mexican government issued 780 postage stamps that read: *Amelia Earhart, Vuelo de buena voluntad Mexico 1935* (Amelia Earhart, flight of good faith, Mexico 1935). Three hundred were sold to the public and fastened to mail that Amelia would carry back with her to America.

Amelia set a new record on her return flight as well. The Vega took off from Mexico City on May 9, 1935, and set down in Newark, New Jersey, in fourteen hours and nineteen minutes. Now she was also the first person to solo from Mexico to Newark!

A flower was named in her honor — the Amelia Earhart Dahlia, a huge blossom in Amelia's favorite color, yellow. Stamps bore her name and image. Countries all over the world honored her. It seemed as though there could not possibly be anything left for Amelia to achieve.

Amelia, however, believed there was much more to accomplish. Travelers of the future, she knew, would be as concerned with the amount of time spent on their flights as they would be with the safety of flying. They would want to fly as quickly as possible. On August 24-25, 1932, Amelia flew from Los Angeles to Newark in nineteen hours, five minutes. Less than a year later she made the same flight, on July 7-8, 1933, and broke her own record with a new time of seventeen hours, seven and one-half minutes.

Purdue University was thrilled with the exploits of their most illustrious faculty member. In late 1935, the Purdue Research Foundation decided to show their appreciation with the gift of a new plane. Amelia chose Lockheed's most modern craft, the Electra 10E, a low-wing, silver-gray plane with two engines and room for ten passengers. It was outfitted with special features, including an autopilot and many other modern instruments. The Lockheed Electra was the most technologically advanced plane of its time, and cost a whopping $42,000. Amelia called it her "Flying Laboratory."

Owning the plane was very important to Amelia's future plans. She had not yet achieved what she believed might be her greatest flight — around the world. The Lockheed Electra was the kind of plane she knew she would need for such a trip. She took delivery of the vehicle on her thirty-ninth birthday, in July 1936. Amelia knew she could now realize even greater triumphs. A flight around the world, impossible in her single-engine Vega, would call attention to the ever-growing horizons of aviation. It would be her biggest and best flight. What Amelia did not know was that it would also be her last.

11

A Flight Around the World

BY THE TIME Amelia received her Lockheed Electra in July 1936, she had already achieved an astonishing number of records in aviation. But in the back of her mind there was still one goal to strive for, one so ambitious she dared not even discuss it in public: Not only did Amelia want to fly around the world, she wanted to fly around the world at the equator.

Others had flown around the world before. U.S. Army air service pilots completed the first round-the-world flight as early as 1924, although the trip took them 175 days. In the early 1930s, Wiley Post, Harold Gatty, and other pilots also circled the globe. However, most flights were planned to circle the world at its narrowest point — usually the North or South Poles. Amelia wanted the challenge of flying around the world at its widest point, the equator. This daring feat had never before been attempted.

Amelia was now thirty-nine years old. Her work, and the work of other pioneering pilots, had helped open the way for the commercial aviation that was beginning to spring up around the world. Air travel had become much more familiar to the public. So much of the earth's airspace had already been covered — a lot of it by Amelia herself — that pioneering flights were beginning to disappear. In just a few years there would be no further need to explore the world by plane. Soon there would be jet speeds that would shatter the sound barrier and lead to the development of rockets that would explore the stars.

Yet Amelia told newsmen she believed she had "one long last flight" in her system. During her flight around the world, she wanted to explore the psychological effects of flight on pilots. This would include factors such as diet, fatigue, mechanical aids, speed, altitude, and pressure. Her silver, twin-engine Lockheed Electra 10E, equipped with every kind of modern instrument and capable of a top speed of 220 miles per hour, was the perfect vessel in which to make this spectacular flight.

In November 1936, Amelia wrote to President and Mrs. Roosevelt, explaining her plans to fly around the world and asking if they would intervene with the Navy on her behalf. The most hazardous aspect of the proposed journey would be the portion over the Pacific Ocean. Amelia wanted to cut down her weight on this portion of the flight as much as possible, which would make maneuvering the Electra far easier and safer. Yet she needed a tremendous amount of fuel to cross over 2,000 miles of ocean to her next stop. She wanted to refuel in the air over the Pacific, in order to avoid carrying heavy fuel loads, and she asked for the Navy's help in accomplishing this. The President responded enthusiastically, and Amelia received complete Navy cooperation.

Amelia turned for technical help to Paul Mantz, who had worked with her on her ground-breaking Pacific flight from Hawaii to California in 1935. Commander Clarence S. Williams, who had set the route for Amelia's Pacific flight, would review her route for the upcoming world flight. Captain Harry Manning, commander of the S.S. *President Roosevelt*, the ship that had carried Amelia home from her first Atlantic flight, signed on as radioman and senior navigator for the flight. He was joined by Fred Noonan, a brilliant navigator who had once worked for Pan American Airways. Noonan had left Pan Am because of a drinking problem. Amelia had faith in Noonan's abilities, however, and trusted in him completely to fly with her throughout the long journey.

Paul Mantz would make sure the Electra was in perfect technical shape, but he would not actually accompany Amelia on the flight. Captain Manning and Fred Noonan would fly with Amelia, working with the radio and navigation while she flew. Manning was particularly valuable because he was the only one of the three who knew Morse code.

When she announced her flight plans to reporters in February 1937, Amelia explained that she would be flying a course from east to west. She and the crew would start in Oakland, California, and fly to Honolulu, Hawaii, where they would refuel. "I think I have just one more long flight in my system," she told newsmen. After the flight was completed, Amelia said, she would enjoy the California sunshine, good books, travel, and friends.

On this trip Amelia intended to carry a little more than she had on her first Atlantic flight. Flying on the *Friendship*, she had limited her personal belongings to simply a toothbrush. Since the flight around the world would take place in the much larger Lockheed Electra, she would carry two suitcases. One would be filled with a change of clothes, the other with maps. Amelia admitted that the map suitcase would be far heavier than the one holding her clothes.

By this time all aircraft were required by law to be registered. Since the Lockheed Electra was highly modified, it needed registration *and* a special license authorization. The license number, NR-16020, showed both of these things. The "N" stood for the United States. Each country had its own letter, which identified planes that originated from it. The "R" stood for "restricted aircraft," and 16020 was the personal number given to Amelia's craft. Every plane also received code letters that were used to identify it over the radio. Amelia's radio-call letters, as the code was known, were KHAQQ.

As a precaution, Amelia had the blue-gray metal of the Electra's wings painted over with black, orange, and red stripes. The bright colors made the plane very noticeable. If she had to make an emergency landing, or if the plane crashed, Amelia wanted to be sure the Lockheed would be seen by passing aircraft or ships.

As usual, G.P. thought of special promotions to bring in money to offset the enormous costs of the flight. It was his idea to have Amelia carry with her 8,000 special envelopes. G.P. called them "stamp covers." These stamp covers would be postmarked at every stop she made on her flight, then released to collectors, who paid for the stamp covers before Amelia began her flight.

G.P. also dreamed up a very dangerous promotional stunt. He asked Amelia to forego the use of a navigator for a portion of the journey, even though she would be flying through thousands of miles of uncharted airspace. He claimed that the publicity would be even better if she flew the dangerous Pacific Ocean leg on her own. Amelia, knowing the expedition was risky even *with* a navigator, said absolutely no. There was no way she would go along with this foolhardy plan.

On March 17, 1937, at 4:37 P.M., Amelia, Fred Noonan, Paul Mantz and Harry Manning took off from Oakland, California, on the first leg of the flight around the world. They reached their first stop, Oahu, Hawaii, in just fifteen hours and forty-seven minutes. It was a new speed record for the trip, as well as an east-west crossing record.

However, Mantz was not pleased with the performance of one of the propellers on the trip. In Oahu, he checked it over, repaired it, and tested it again before giving Amelia the go-ahead to fly out of Hawaii. Mantz was staying in Hawaii, while Manning and Noonan were continuing on with Amelia.

By the time Mantz was satisfied with the propeller,

however, the Hawaiian weather had turned sour, forcing the crew to wait until March 20 before they could set off from Hawaii. The next leg of the flight would be the most difficult on the journey: They were aiming for tiny Howland Island, a dot of land in the Pacific Ocean. Even with all of the Electra's equipment, it would be a challenge to arrive at Howland successfully. Both Manning and Noonan would be needed for navigation if they were to hit the island precisely.

Early on the morning of March 20, Amelia taxied down the runway at Oahu, but the plane did not respond correctly. Heavy and overloaded with fuel, the Electra pulled down and to the right, resulting in a "ground loop" at the end of the runway. A ground loop is a spin on the ground that occurs when the plane tips too much to one side. In this case, the Electra's landing gear was already up, in position for takeoff. The plane simply spun slowly on its belly, tearing away large pieces of the wing and underside.

Amelia, cool in any emergency, immediately cut the engine to avoid the possibility of a fire. But as she climbed out of the Electra and looked at the damaged plane, tears rose to her eyes. Her trip had come to a halt almost before it began! How would she ever be able to afford to repair the plane and start again?

She had no choice. The Electra went into the Lockheed factory for repairs, while Amelia and G.P. tried to raise more money for a second attempt. Many people, admiring Amelia's courage, stepped forward with donations. Generous donors included such well-known figures as Jacqueline Cochran, the flier, and millionaire statesman Bernard Baruch, who wired $2,500 to Amelia, "because I like your everlasting guts!" The money continued to come in as the plane was being repaired and refitted.

Finally, on May 17, the Electra was fit to fly again. Amelia wired Dr. Elliott at Purdue University: *Our second attempt is assured. We are solvent. Future is mortgaged, but what else are futures for?*

The two months' delay changed a number of important factors in the flight. For one thing, Captain Manning was unable to accompany Amelia and Fred Noonan, as he had planned. He had taken a leave of absence from his ship for the flight. By the time the Electra was ready to fly again, so much time had passed that Manning was forced to report back to his ship. This meant that the sole navigator on the flight would be Noonan. Amelia would do her best with the radio, but neither she nor Noonan had the expertise in radio communications that Manning had.

In addition, heavy Pacific storms were blowing up, as they did every year in the late spring. If Amelia had flown out on schedule, she would already have put the storms behind her. Now, however, she would be flying right into them. It was necessary to reroute the flight to avoid the bad weather. Instead of flying east to west, they would now fly west to east. Amelia also decided to eliminate entirely her planned stops in Central America, fearing that the heavy tropical spring rains there might delay her.

The new plan also called for a cautious testing of the Electra. Amelia would fly the plane from Oakland, California, to El Paso, Texas, and then on to

Miami, Florida. These were considered safe flights in familiar, well-known airspace. Should anything go wrong, it would be easy to quietly return to the Lockheed plant for further repairs. If the plane was all right, the flight would proceed.

Before announcing to the press the new plans for her revised trip, Amelia made the first test flights with Fred Noonan. They arrived in Miami confident that the plane was in top condition. At last, on June 1, 1937, Amelia and Fred took off from Miami on their flight around the equator.

The entire world eagerly kept track of the flight. Amelia sent regular dispatches to G.P., which he gave to a newspaper syndicate for publication. Every day, it seemed as though the Electra was in the news as it sped around the world, encountering curiosity, admiration, and hearty welcome.

When the crew had time to stop, they often shopped and called home, as other tourists do. Amelia bought some pretty, brightly colored bracelets, which she mailed to her young niece, Amy. She saw a lovely sari, or Indian dress cloth, she wanted to buy for herself, but she decided against it. The Electra was flying under strict weight conditions. Absolutely no excess baggage was permitted.

As the journey moved eastward, the air grew hotter and more humid, and Amelia and Fred grew more and more tired. Several times, Fred spent his off-nights drinking with the mechanics. Amelia handled his drinking problem with her typical caring and helpfulness. Each time he drank, she sought him out, helped to sober him up, and got him ready to fly the next day. It is not surprising that Fred wrote to his wife that Amelia was a wonderful companion to have on the long flight.

The Electra flew down the coast of South America and turned east toward Africa. The plane then cut over to Karachi, India, heading down to Singapore and across to Darwin, Australia, and Lae, New Guinea. They reached Lae on June 29.

During their stop at Lae, as on the other stops, the Electra needed maintenance and refueling. Amelia and Fred caught up on sleep and planned the next leg of the flight, the most dangerous of all. They would cross through uncharted airspace over 2,000 miles of ocean to Howland Island. Once they had refueled on Howland, they would fly on to Hawaii, and from there to Oakland, California, to complete their trip around the world. Amelia expected to arrive back in the United States around July 4.

The weather on Lae soon changed her plans. For several days, the Electra crew was hemmed in by bad weather. Finally, on July 2, Amelia received clearance to fly on. The U.S. Department of the Interior had taken over from the Navy the job of helping Amelia during this dangerous part of the trip. They had built a runway at Howland Island. Everything was ready for Amelia to fly in.

The U.S. Coast Guard supplied extra help at Howland Island. A cutter called the *Itasca* was given the job of guiding the Electra to the island. To do this, they would communicate with the Electra by radio, and use black smoke signals as a landmark that Amelia and Fred would be able to see from the plane.

Unfortunately, the crew of the *Itasca* could not quite figure out what Amelia wanted to use in the way of radio frequencies. Communication between the Electra and the *Itasca* was extremely important. The *Itasca*'s crew members expected Amelia to contact them on strong radio frequencies so they could pick up her signals easily. Instead, she requested frequencies whose signals were weak and hard to read. The Itasca could not understand why.

The Coast Guard sent a variety of messages requesting that Amelia use different, stronger radio frequencies. Amelia never responded to their queries. Since she had the final word on this flight, the *Itasca* obediently agreed to use the weaker radio frequencies Amelia wanted.

At ten o'clock in the morning of July 2, Amelia and Fred took off from Lae, flying east toward Howland Island. They expected to arrive eighteen hours later. Early on the morning of July 3, close to their destination, the Electra crossed the international date line. This meant a time change of twenty-four hours. In other words, because they were flying from west to east, they were flying backward in time. It was July 2 again, very early in the morning.

Radio operators were standing by on Howland Island as well as on the *Itasca*. Amelia reported that she would use frequency 3105 and would report in on the hour and on the half hour. Yet the crew of the *Itasca* had a hard time reading her signals. They simply could not keep her on the air. Nor could they communicate in Morse code, since neither Amelia nor Fred Noonan understood it. They begged her to change radio frequencies, but Amelia did not seem to hear them.

At 4:53 A.M., the *Itasca* radioman heard Amelia say "partly cloudy." A few minutes later she asked for a bearing and said she would whistle into the radio microphone. They heard her say she was 200 miles out. Again she asked them to take a bearing on the plane. The *Itasca* tried, but Amelia did not stay on the radio long enough for them to do so.

The situation grew desperate. The *Itasca* repeatedly tried to reach Amelia, broadcasting messages to her over and over again. She answered none of them. It was as though she had not even heard them. Finally she sent a message that chilled them: "We must be on you but cannot see you. Gas is running low. Been unable to reach you by radio. Flying at 1,000 feet. Only half hour's gas left." By this time, the *Itasca* was sending up thick black smoke to guide Amelia and Fred to the island. It was clear overhead, and it seemed impossible to the ship's crew that the Electra failed to see the smoke.

As morning broke over the Pacific, Amelia broadcast again: "We are in line of position 157-337. Will repeat this message on 6210 KC [another radio frequency]. Wait listening on 6210. We are running north and south."

The *Itasca* waited and waited. They broadcast again, hoping to hear from Amelia, but there was no further broadcast from the plane. The Electra, Fred Noonan, and Amelia Earhart vanished without a trace.

12

What Happened to Amelia?

FROM THE MOMENT of her last radio transmission to the *Itasca*, Amelia, Fred Noonan, and the Lockheed Electra vanished without a trace. The *Itasca* radioed again and again, and its radio operators waited hopefully for another transmission from Amelia. There was none.

What could have happened?

Amelia had radioed that she had only thirty minutes of fuel remaining in her fuel tanks. If the *Itasca* did not hear from her, its commander estimated that she would be forced to crash-land in the ocean when her fuel tanks ran dry. With this in mind, the *Itasca* began a search for the plane within a few hours of her last radio transmission.

Over the next few days, the Coast Guard and Navy sent out ships and planes in a massive search for the downed Electra. Hundreds of sailors and pilots were involved in the search. Amelia's close friend Jacqueline Cochran, a famous aviator herself, volunteered her services. Cochran flew hundreds of miles over the Pacific Ocean looking for Amelia.

The effort was futile. After twelve days of the most intensive manhunt ever launched by the government, the official search for Amelia and Fred ended. The United States had spent four million dollars to look for the aviator and her navigator, without results. Amelia's family began its own search, but they learned nothing. The situation was baffling. One of the twentieth century's greatest mysteries had begun.

Amelia's husband, G.P. Putnam, believed that Amelia and Fred were alive. He continued to badger government officials for information and assistance long after their disappearance. As time passed, however, hope began to diminish. World War II had begun in Europe, and the United States was facing the possibility of entering the war. The country had little attention to spare for the fate of Lady Lindy.

In January 1939, seventeen months after the Electra's disappearance, G.P. Putnam appeared in court to request that his wife, Amelia Earhart, be declared officially dead. With no word and no hope left, he wanted to go on with his life. The court declared that Amelia was officially dead and probated her will, which left her entire estate, valued at $27,000, to her husband. According to the United States court, G.P. Putnam was now a widower. He married another woman four months later.

Unlike G.P., Amelia's mother and sister continued to pray that Amelia was alive and that she would reappear when the war was over. But when World War II ended, in 1945, Amelia did not reappear.

What really happened to Amelia Earhart and Fred Noonan on that July morning of 1937? Over the years, many questions about their disappearance have been raised. The possible answers are complex, dramatic, and fascinating:

Did Amelia and Fred crash in the Pacific Ocean?

This was the government's first theory. Investigators reasoned that the Electra's fuel tanks ran dry, and that, unable to spot land, Amelia was forced to come down in the ocean. This theory presumes that Amelia and Fred died when the plane hit the water, and that the Electra disintegrated in the ocean.

Some doubt has been cast on this theory. A plane as large as Amelia's would not sink immediately upon hitting the water. In addition, if it actually broke apart as the investigators believed, much of the plane's fluid would have been exposed on the water. The government launched its search for the fliers within a few hours after their last radio transmission. Planes were flying over the area almost immediately afterward, and ships followed, scanning for any trace of the Electra. Neither the ships nor the planes picked up any trace of a large oil slick, which would almost certainly have trailed off from the wreckage.

It is possible that Amelia was able to set the plane down in the ocean without damaging it. The mechanics who knew the Electra said that if that had

The Bettman Archive

Earhart vanished in July 1937 during her attempt to fly around the globe at the equator. Did the plane run out of fuel and crash? Was Amelia on a spying mission? Her last flight sparked one of the greatest, and still unsolved, mysteries of the twentieth century.

happened, the plane would have been able to float for an indefinite period of time. The empty fuel tanks would have provided flotation for several hours, if not a full day. Surely someone flying over such a large area or scanning on a ship would have spotted them.

Did Amelia and Fred crash-land near an island?

With no fuel and unable to see the *Itasca*'s smoke signals, it is perfectly reasonable that Amelia might have landed the plane in the ocean, as close to land as she could manage.

For many days after the disappearance of the Electra, ham radio operators in the United States reported hearing strange messages from a woman using the Electra's radio call letters, KHAQQ. They reported pleas for help from the flier, who said she was stranded on an unnamed island.

One particularly interested fifteen-year-old ham-radio buff, Robert Myers of Oakland, California, heard a message that haunted him. He did not reveal the incident at the time, because he felt that no one would have believed a fifteen-year-old boy. With so many other people claiming to have made contact with Amelia, Myers felt he would be ridiculed. He waited over forty years before coming forward with his story.

While fiddling with his radio set, Myers heard a woman who claimed to be Amelia saying that her plane had used up all its fuel searching for Howland Island. The woman said she had crash-landed near

Hull Island in the Pacific. Her navigator, Fred Noonan, had hurt his head in the crash.

As the boy listened, the woman's voice became frightened. She said that Japanese soldiers had entered the plane and were beating Fred Noonan. Moments later, the boy heard male voices speaking in a foreign language. He heard the woman cry out, asking them not to hurt her. The transmission went dead. Myers never heard another word from the mysterious woman.

There are also reports of messages written on scraps of paper and stuffed into bottles, which were retrieved by people in various parts of the world. A Frenchwoman found such a bottle floating at the water's edge near her home. The message, written in lipstick on a piece of paper, said that the writer was Amelia Earhart and that she and her navigator were on an uninhabited island somewhere in the Pacific. This was the last paper she had. She begged the finder of the note to please contact the United States government and her husband, G.P. Putnam.

In 1960, in San Mateo, California, a woman named Mrs. Josephine Blanco Akiyama claimed that she had seen Amelia and Fred on the island of Saipan in the Mariana Islands in 1937. As an eleven-year-old girl on Saipan, Josephine was taking lunch to her brother-in-law, who worked in a secret Japanese seaplane base at Tanapag Harbor on the western shore of Saipan. She saw a large twin-engine plane fly overhead and out of sight, in the direction of the harbor.

When Josephine arrived at the harbor, she found a group of people gathered around two thin white people who looked very tired. At first she thought they were both men, but someone told her that one was a woman. The woman's hair was cut short, and she wore men's clothes. The man's head was injured. Japanese guards took the pair away from the harbor. Josephine later heard rumors that the white people were executed by the Japanese as spies.

Saipan was controlled by the Japanese in 1937, along with the Marshall Islands. According to the rules of the League of Nations, the forerunner of today's United Nations, the Japanese were forbidden to fortify the islands or develop their military strength. Despite the rules, Japan was secretly building military fortifications in the Marshalls. They intended to be ready for war when it came. Because the Japanese were violating the League of Nations' rules, they could not risk permitting foreigners on the islands. When Amelia and Fred disappeared, the Japanese did not permit American search planes to fly overhead looking for the missing aviator, and they did not permit American ships in their harbors.

When the Americans stepped up their search for Amelia and Fred, the Japanese made a curious offer to the U.S. government. They told the government they would conduct their own search for Amelia within Japanese territory. They had ships which could scan the area for her, and they would report to the U.S. armed forces if they found the downed fliers. As the government search for Amelia and Fred began to wind down, twelve days after their disappearance, the Japanese reported that they had not found a thing. They said they hoped the Americans had had better luck.

Every indication from those last radio transmissions suggests that the Electra was flying above Japanese territory. Amelia reported heavy clouds in her flight path. The weather around Howland Island was entirely clear — except in the northwest. To the northwest lay heavy clouds, and the Marshall and Mariana Islands in Japanese hands.

The *Itasca* had been instructed to send up black smoke to guide Amelia toward Howland Island. But Amelia radioed that she could not see the smoke and was in heavy clouds. It is almost certain that she was flying northwest of Howland. Possibly she did not realize how far off-target she was.

Were Amelia and Fred taken prisoner by the Japanese?

Over the years, several different American investigators have tried to solve the Earhart mystery. A number of them visited the islands in the Pacific looking for witnesses who might remember seeing the fliers in 1937.

All of the researchers, without exception, found natives on Saipan who spoke of having seen a "white woman and white man" arrive in "a big sky bird" a few years before the war began. Since World War II began in 1939 and Amelia and Fred disappeared in 1937, this description would fit them perfectly.

One persistent investigator, a radio newsman named Fred Goerner, showed photographs of different women, including Amelia, to natives who claimed to remember the white woman. Each native without hesitation picked out Amelia's photograph as the woman they remembered.

In 1937, Saipan was the headquarters for the Japanese military efforts in the Pacific. Anyone arriving unexpectedly anywhere in the Marshall or Mariana islands would probably have been taken to Saipan for questioning by the Japanese. It is likely that if the Japanese believed they were spies, they would have been held indefinitely.

There is a jail on Saipan which was used for prisoners of the Japanese during World War II. Some Saipan natives claim to have seen Amelia there. They said that she was interned on Saipan for extensive questioning. According to those reports, the Japanese did not believe that Amelia was lost. They believed that she was trying to spy on their military operations.

One native woman remembered that it was her job to do Amelia's laundry. Her story, which was told to newsman Fred Goerner, claims that the American woman became tired, sad, and ill. Under guard, she walked around an enclosed area in the jail every day for exercise. One day the native woman saw Amelia walking there. She said that the white woman smiled at her and patted her head, as if to thank her.

A number of Saipan natives remember seeing Amelia and Fred in the years before the war. It seems reasonable to assume that the Electra crash-landed in Japanese territory — whether on Saipan or another island — and were imprisoned on Saipan for some time.

Did Amelia and Fred Noonan survive World War II?

American troops landed on Saipan in July 1944. They had ample opportunity to search the island. A small detachment of Marines found a room fixed up for a woman, as well as various articles of women's clothing. They also found a diary on which was stamped the words, *Ten-Year Diary of Amelia Earhart*. They gave these items to their commanding officer, who promised to forward them to the proper authorities. The articles disappeared completely.

Other puzzling events took place. Two Marine privates joined an officer on a mysterious task. The officer asked them to spend a day excavating at a certain spot on Saipan. When one of the Marines asked what they were digging for, the officer answered, "Have you ever heard of Amelia Earhart?" The Marines eventually found a hole full of bones, which the officer stored in metal canisters.

A military man named Joe Gervais pursued the theory that Amelia lived through World War II. He believed that she might have been the voice of the infamous "Tokyo Rose." Tokyo Rose was a woman who spoke on the radio to American G.I.'s, giving them false information in order to discourage them from fighting at their best. (American servicemen were often referred to as G.I.'s, which stands for government issue.) However, others say that it doesn't make sense that Amelia would try to undermine the U.S. war effort in World War II. One explanation is that the Japanese, while keeping her in their custody, brainwashed Amelia into turning against her native country.

G.P. Putnam, who had an army commission during the war, made a special trip to hear the voice of Tokyo Rose, the woman he thought might be his wife. After listening for a minute, he insisted it was not Amelia. He said the woman had been well coached and might have come from the New York area, but it was definitely not his wife.

Amelia's mother and sister hoped that once the war was over, Amelia would return home. They reasoned that if Amelia had been held prisoner by the Japanese, their defeat would mean her release. They believed that if she was alive she would be coming home — even if she had to swim all the way.

There is far less confusion about whether Fred Noonan survived the war. From almost all reports, Fred angered his Japanese captors and was executed shortly after he and Amelia were taken prisoner.

Were Amelia and Fred actually spying for the U.S. government?

In 1943, RKO Pictures released an adventure film called *Flight for Freedom*. The story, though fictional, drew heavily from the true story of Amelia's last flight. The film was about a courageous female aviator flying around the world with a male navigator with whom she was in love. In the film, the aviator had been approached by the U.S. government to help them learn about secret Japanese fortifications on isolated Pacific islands.

The aviator in the movie agreed to photograph the islands from the air. When she realized she might not be able to deliver the photographs, she decided to give the government a reason to launch a search for her and take their own pictures. With great courage, the screen heroine deliberately crashed her plane at sea, in order to inspire a hunt that would double as a photographic mission for the U.S. military. She knew even before she crashed that she and her navigator might die without being rescued.

This was the first indication to the public that Amelia might have been on a special mission that was kept very secret. Fred Goerner, in his research, sought out Amelia's secretary, Margot DeCarrie. DeCarrie told Goerner that Amelia had entrusted her with a large sealed manila envelope just before she started her world flight. Amelia had told her never to open it. She said that she could not trust her husband or her mother to obey such instructions, so she was giving the envelope to Margot. When she returned, Margot would hand over the envelope, sealed. In the event she did not return, Amelia instructed Margot to burn the envelope without opening it. DeCarrie told Goerner that she had never opened the envelope. When Amelia was lost, DeCarrie said, she burned it in accordance with Amelia's instructions.

In addition to the peculiar instructions given to DeCarrie, there are other indications that tend to support the theory that Amelia had special plans for her last flight beyond simply circling the globe.

Amelia made a curious, off-the-record comment to a newsman, Carl Allen, just before she climbed aboard the Electra to take off on the first leg of the flight around the world. "You know," she said to Allen, "I am not coming back from this flight." Allen agreed to keep her remark confidential, though at the time he vigorously disagreed with her.

Amy Earhart admitted that though she was very close to her daughter, Amelia told her a great deal but did not tell her everything. There were some things, Amy said candidly, that she knew Amelia could not tell her about that last flight.

Amelia was a good friend of President and Mrs. Roosevelt. It is not impossible that the President might have asked for her help. If she was spying for the President, it is hardly surprising that Amelia would keep quiet about it. For one thing, she would keep such preparations secret for national security reasons. Also, in 1937, spying for one's country was thought to be a lowly occupation.

Consider the fact that Amelia was granted government cooperation from the military for what was, unquestionably, a civilian flight. The military supported the Electra with ships and personnel. They also built a special runway for the plane at Howland Island. In addition, when the Electra became lost, the military launched a four-million-dollar search for the plane. The twelve-day manhunt was the most extensive ever conducted, encompassing 25,000 miles of ocean. Would the government really have spent so much money looking for a mere civilian on a daredevil flight? Other civilian pilots, seeking help for their own missions, had been rudely refused by the government. Even some of the smallest requests had been turned down. Why was Amelia given so much help? Wasn't she just a civilian, too?

Over the years, researchers have puzzled over the radio transmissions received from Amelia on the morning of her disappearance. Neither she nor Fred Noonan knew the International Morse code, which is used for radio communication around the world. Even more puzzling is the fact that they had left some important radio equipment in Miami, Florida, rather than carry it with them on the flight.

Amelia knew that her flight over the Pacific would be dangerous, since it had never been done before. She needed the guidance of the *Itasca*. The clear and continuous radio contact the ship provided was vital to the success of her flight. Yet she refused to broadcast on a frequency that would be easily readable. The *Itasca*'s suggestions on which frequencies to use went unanswered.

Paul Rafford, a former pilot, theorized that Amelia did not *want* the *Itasca* to be able to pinpoint exactly where the Electra was. This is because, he believed, she was on a spy mission that was to take the Electra far out of its intended flight path in order to photograph Japanese operations. When that was accomplished, she would maneuver the Electra to Howland Island.

At almost any point during the flight, the *Itasca* could have pinpointed Amelia's location if Amelia had simply held down her radio transmission button for several seconds. However, all through her communication with the *Itasca*, she made brief, hurried transmissions and got off the air very swiftly. She also switched to different frequencies. If she and Fred were spying, it makes sense that she would remain off the air for long periods of time, make only the briefest transmissions, and change frequencies. This had to confuse the *Itasca*, while giving the impression that Amelia was simply lost and unable to understand how to use her equipment properly.

Does the government really know the outcome of Amelia's last flight but refuse to make public that information?

Researchers through the years have often suggested that the U.S. government has a secret file revealing exactly what happened to Amelia and Fred Noonan. Fred Goerner, during his six-year struggle to unravel the mystery, found himself hitting one stone wall of silence after another in the government.

Admiral Chester Nimitz, the highest-ranking man in the U.S. Navy at the time of Amelia's disappearance, was one of the few officials who encouraged Goerner to continue his research. Nimitz said that he could not reveal to Goerner what he himself knew, but he remarked that the truth would astonish him. Nimitz also confirmed that Amelia and Fred had crash-landed in the Marshall Islands and been picked up by the Japanese.

With the passing of the Freedom of Information Act in 1966, many formerly classified documents became available for review. These include the logs of the *Itasca* with copies of the radio messages sent and received on the morning of the Electra's disappearance, and correspondence with the Japanese government.

Henry Morgenthau, President Roosevelt's secretary of the treasury, had a telephone conversation with Mrs. Roosevelt's secretary on May 13, 1938. In this conversation Morgenthau indicated that he knew a great deal about what had happened to Amelia. He stated that if the government were to release some of this information, it could "smear the whole reputation of Amelia Earhart." Was Morgenthau referring to her inability as a pilot to pick up the *Itasca*'s radio signals? Or perhaps to her secret position as spy for the government?

There is still no answer to those questions. While there may be classified files in existence that provide information about Amelia, those files that are on public record provide no real answer to what happened to her.

Is Amelia still living under an assumed name?

In 1970, writer Joe Klaas published a book called *Amelia Earhart Lives*. The book is an account of the attempts of Joe Gervais, a former Air Force pilot, to track down a woman he believed was Amelia Earhart. Gervais, who had over 8,000 flying hours to his credit, had appeared at a luncheon on Long Island in 1965 for the Early Fliers' Club. As he chatted with some of the women who had made aviation history — many of whom had known and flown with Amelia — he saw a woman in the doorway. As he stared at her, he realized her features were very familiar. The woman in the doorway resembled Amelia so strongly she could almost have been her twin. Furthermore, she wore a tiny red, white, and blue ribbon — which can only be worn by winners of the Distinguished Flying Cross (DFC). Amelia was awarded the DFC after her Atlantic solo in 1932.

Gervais maneuvered his way into a conversation with the woman, who introduced herself as Irene Bolam, the wife of Guy Bolam, a financial investor. She admitted to having known Amelia and to having flown with her. In answer to Gervais's questions, she said yes, she was a Ninety-Nine and a member of Zonta International. Amelia had belonged to both organizations. When Gervais later tried to look up Mrs. Bolam's memberships, neither organization had her listed as a member!

Mrs. Bolam explained that she was granted her license under her previous name. But Gervais found it curious that she became licensed only a few months before Amelia set off on her flight around the world in 1937. He wondered whether this was deliberate. Was it possible that Amelia had taken a license in a new name in order to continue to fly under a new identity when she returned to the United States after her spying mission?

Mrs. Bolam was very mysterious and very secretive. Few people who lived nearby even knew who she was. The house in which she lived belonged to Amelia's close friend Jacqueline Cochran and Jacqueline's husband, Floyd Odlum.

Mrs. Bolam evaded all of Gervais's attempts to talk to her seriously about her identity. In 1970, when the book about his search was published, she went to court to demand that the book be withdrawn. She insisted she was not Amelia Earhart.

Gervais believed that the price of Amelia's life was very great for the U.S. government. America was not prepared to go to war in 1937. Therefore, the President could not press the Japanese government for a definitive answer about Amelia. It is possible that she was a hostage of the Japanese, and that the American government agreed to certain demands by the Japanese in order to buy time before they entered the war. Once the U.S. military strength was at its peak, they could then try to free Amelia by force, if necessary.

In exchange for keeping Amelia alive, Gervais believed the U.S. might have agreed to give the Japanese the plans for the Zero plane. The Zero plane proved effective for the Japanese in combat during World War II. If such a deal to keep Amelia alive *was* made, it could also be the reason President Truman agreed not to prosecute Japan's Emperor Hirohito as a war criminal when the war finally ended in 1945.

What really happened to Amelia Earhart?

The answers might never be revealed. She might have crashed in the ocean or been captured by the Japanese. She might have been murdered, or she might still be living quietly in some American town. The truth may never come to light. But the fact remains that the woman of smiling courage who stepped into her Lockheed Electra in early July 1937 has vanished forever.

13

Amelia's Achievements

AMELIA EARHART'S CAREER in aviation spanned a mere fifteen years. She was internationally famous for only ten of those years. Yet her legacy to flying and to women who dream of reaching new horizons endures today. She was a pioneer who believed in continually testing, and crossing, new boundaries.

Amelia set a marvelous example for women everywhere. She believed firmly that women must always seek new challenges independent of men, that women's lives are just as important and should be as adventurous as men's lives. Again and again she repeated that her most important goal was encouraging women to take to the air.

Her own life vividly illustrates Amelia's conviction, courage, and independence. Daily, she risked everything in the rush of exploration that characterized her life. She constantly set new records and achieved what others insisted women could never achieve.

Here is a list of Amelia's major aviation records:

October 1922

* Women's altitude record — 14,000 feet

June 1928

* First woman to cross the Atlantic Ocean by air

July 1930

* Speed records for 100 kilometers and for 100 kilometers with a freight of 500 pounds

April 1931

* Altitude record for autogiros — 15,000 feet
* New altitude record for autogiros — 18,415 feet
* First woman to pilot an autogiro cross-country

May 1932

* First woman to cross the Atlantic solo
* First person to cross the Atlantic Ocean twice

August 1932

* Speed record for women's nonstop trancontinental flight — Los Angeles, California, to Newark, New Jersey, in 19 hours, 5 minutes

July 1933

* New record for transcontinental travel, breaking her own existing record — 17 hours, 7 ½ minutes

January 1935

* First person to fly from Honolulu, Hawaii, to Oakland, California — 2,408 miles in 17 hours, 7 minutes
* First person to solo anywhere in the Pacific
* First person to solo over both the Atlantic and Pacific Oceans

April 1935

* First person to solo from Los Angeles, California,

to Mexico City, Mexico — 13 hours, 23 minutes

May 1935

* First person to solo from Mexico City, Mexico, to Newark, New Jersey — 14 hours, 19 minutes

March 1937

* Record for east-to-west crossing — Oakland, California, to Honolulu, Hawaii, in 15 hours, 43 minutes

June-July 1937

* First person to attempt to fly around the globe at the equator — 22,000 miles flown, journey incomplete

Even on her last flight, Amelia accomplished important strides in aviation. Because of her disappearance, future flights were required to adhere to a set of basic emergency procedures. All dates are now required to be given in Greenwich mean time, the international standard, instead of the time for the zone in which a flier is stranded.

Macon Reed of the *Navy Times* noted that, "Amelia Earhart flew to advance the interests of aviation. That was her mission. She may not have accomplished what she set out to do in this last flight, but advance aviation she did."

The mystery surrounding Amelia's last flight has attracted legions of investigators and other nonfliers. As they read the details of that final tragedy, the curious everywhere find themselves drawn not only to Amelia's mystery, but to her wonderful spirit.

With the passing years, hundreds of tributes to America's premier woman flier have been heard from every corner of the country. Statues of Amelia, postage stamps featuring her face, street signs and luggage bearing her name, and even a mountain peak named after her in California's Yosemite National Park, keep her in the public consciousness. Poets, writers, and artists have given voice to their feelings about her.

Yet the comment that best sums up Amelia's life came not from a poet but from a man in the U.S. Navy. His job was to report on the *Itasca*'s search for Amelia and Fred around Howland Island. The Navy man said simply, "However she came down, Amelia was not a failure."

Other books you might enjoy reading

1. Brennan, T.C. Buddy. *Witness to the Execution/ The Odyssey of Amelia Earhart.* Renaissance House, 1988.

2. Devine, Thomas E. with Richard M. Daley. *Eyewitness: The Amelia Earhart Incident.* Renaissance House, 1987.

3. Goerner, Fred. *The Search for Amelia Earhart.* Doubleday & Co., 1966.

4. Klaas, Joe. *Amelia Earhart Lives.* McGraw-Hill, 1970.

5. Loomis, Vincent. *Amelia Earhart: The Final Story.* Random House, 1985.

6. Morrissey, Muriel Earhart. *Courage Is the Price.* McCormick-Armstrong Publishing Division, 1963.

7. Parlin, John. *Amelia Earhart.* Dell Publishing Co., 1962.

8. Southern, Neta Snook. *I Taught Amelia To Fly.* Vantage Press, 1974.

9. Strippel, Dick. *Amelia Earhart, The Myth and the Reality.* Exposition Press, 1972.

ABOUT THE AUTHOR

Susan Sloate has written several screenplays and is the author of *Abraham Lincoln: The Freedom President* in the Great Lives Series. She contributes sports interviews to *Baseball Card News*, a collector's publication. In her spare time, she rides horses, pilots airplanes, and goes white-water rafting.